Gun Violence

Gun violence is a vexing problem in the US that needs little explanation. Yet adequate information and insight are still lacking about ownership and use of firearms, the causes and consequences of gun violence, and the effects of interventions and technological innovations. The sociological context and the implications for health and law enforcement involve a variety of disciplines; and the spectrum of questions that arise affects policymakers at all levels. Obtaining adequate information and understanding requires high quality data and analyses that can withstand scrutiny by all concerned.

This book provides an overview of reliable data and informative analyses on several important topics in gun violence research. It exposes the opportunities and needs for cutting-edge statistical efforts and engages new statisticians and data scientists to work across disciplinary lines to help understand and mitigate the effects of gun violence.

Key Features:

- Provides multi-disciplinary perspectives from researchers in criminology, political science, public policy, public health, sociology, and statistics, presenting their latest efforts in gun violence research across disciplinary lines.
- Provides an overview of the state-of-the-art statistical and machine learning methods in gun violence research, including network models, synthetic control methods, point processes models, and neural networks.
- Provides a comprehensive review of data sources, describes the evolution of data quality, and examines the reasons for biases historical data sources.

Gun Violence

The Statistical Issues

Edited by
Charles Loeffler, James L. Rosenberger, and
Lingzhou Xue

CRC Press
Taylor & Francis Group
Boca Raton London New York

CRC Press is an imprint of the
Taylor & Francis Group, an **informa** business
A CHAPMAN & HALL BOOK

Designed cover image: Andrew Holbrook

First edition published 2027
by CRC Press
2385 NW Executive Center Drive, Suite 320, Boca Raton FL 33431

and by CRC Press
4 Park Square, Milton Park, Abingdon, Oxon, OX14 4RN

CRC Press is an imprint of Taylor & Francis Group, LLC

ISBN: 9781032554495 (hbk)
ISBN: 9781032554501 (pbk)
ISBN: 9781003430773 (ebk)

DOI: 10.1201/9781003430773

Typeset in Palatino LT Std
by Deanta Global Publishing Services, Chennai, India

Contents

Editor Biographies

Charles Loeffler is Associate Professor of Criminology at the University of Pennsylvania, and his research interests include life-course criminology, the effects of criminal justice institutions, and topics at the intersection of crime and technology. He was a First Place Winner of the 2017 National Institute of Justice (NIJ) Real Time Crime Forecasting Competition and was a National Science Foundation/National Bureau of Economic Research Crime Research Fellow. He was the co-organizer of the 2024 Ingram Olkin *Statistics Serving Society* Forum on "The Statistics of Gun Violence."

James L. Rosenberger is Professor Emeritus of Statistics at the Pennsylvania State University and the Director Emeritus of the National Institute of Statistical Sciences (NISS), and his research interests include linear models, design and analysis of experiments, bioinformatics, and genomics. He is a Fellow of the American Statistical Association (ASA) and a Fellow of the American Association for the Advancement of Science (AAAS). He was the co-organizer of the Inaugural Ingram Olkin *Statistics Serving Society* Forum on "Gun Violence: The Statistical Issues" in 2019 and the 2024 Ingram Olkin *Statistics Serving Society* Forum on "The Statistics of Gun Violence."

Lingzhou Xue is Professor of Statistics at the Pennsylvania State University, and his research interests include high-dimensional statistics, statistical and machine learning, nonparametric statistics, optimization, and statistical applications in biomedical, environmental, and social sciences. He is a Fellow of the Institute of Mathematical Statistics (IMS), a Fellow of the American Statistical Association (ASA), and an Elected Member of the International Statistical Institute (ISI). He was the co-organizer of the Inaugural Ingram Olkin *Statistics Serving Society* Forum on "Gun Violence: The Statistical Issues" in 2019 and the 2024 Ingram Olkin *Statistics Serving Society* Forum on "The Statistics of Gun Violence."

Contributors

Ian Adams, PhD, is Assistant Professor of Criminology and Criminal Justice at the University of South Carolina.

Catherine Barber is a senior researcher in the Harvard Injury Control Research Center at the Harvard T.H. Chan School of Public Health.

Eli Ben-Michael, PhD, is Assistant Professor in the Department of Statistics & Data Science and the Heinz College of Information Systems and Public Policy at Carnegie Mellon University.

P. Jeffrey Brantingham, PhD, is Professor of Anthropology at the University of California Los Angeles.

Ryan Cecil is a PhD candidate in the Department of Statistics at the University of Pittsburgh.

Philip J. Cook, PhD, is Professor Emeritus of Public Policy and Economics at Duke University.

Zheng Dong, PhD, is an applied scientist at Amazon.

Mitchell L. Doucette, PhD, is a researcher at the Johns Hopkins Bloomberg School of Public Health.

Avi Feller, PhD, is Associate Professor of Public Policy and Statistics at the University of California, Berkeley.

Andrew J. Holbrook, PhD, is Associate Professor of Biostatistics at the University of California, Los Angeles.

Alex Knorre, PhD, is Assistant Professor in the Department of Criminology at the University of South Florida.

Seyoon Ko, PhD, is Assistant Adjunct Professor of Mathematics and Biostatistics at the University of California, Los Angeles.

Daniel S. Lawrence, PhD, is a senior research scientist at the CNA Corporation.

Charles E. Loeffler, PhD, is Associate Professor of Criminology at the University of Pennsylvania.

Alexander D. McCourt, PhD, is Assistant Professor at the Johns Hopkins Bloomberg School of Public Health.

Josh McCrain, PhD, is Assistant Professor of Political Science at the University of Utah.

Lucas Mentch, PhD, is Associate Professor in the Department of Statistics at the University of Pittsburgh.

George Mohler, PhD, is Professor of Computer Science at Boston College.

Fred Morstatter, PhD, is Research Associate Professor at the University of Southern California.

Justin Nix, PhD, is Distinguished Professor of Criminology and Criminal Justice at the University of Nebraska at Omaha.

Sarah Oliver holds a BA in Computer Science from Boston College and is a software engineer at Amazon.

Susan T. Parker, PhD, is a research assistant professor in the Department of Emergency Medicine at Northwestern University.

Sarah Penrice holds a BS in Computer Science from Boston College and is a network design engineer at Verizon.

Eric L. Piza, PhD, is the Lipman Family Professor of Criminology & Criminal Justice and Director of Crime Analysis Initiatives at Northeastern University.

James L. Rosenberger, PhD, is Professor Emeritus of Statistics at The Pennsylvania State University.

Elizabeth A. Stuart, PhD, is the Frank Hurley and Catharine Dorrier Professor in the Department of Biostatistics at the Johns Hopkins Bloomberg School of Public Health.

Audrey Vila, JD, is a 2026 graduate of Boston College Law School.

Yao Xie, PhD, is the Coca-Cola Foundation Chair and Professor at the H. Milton Stewart School of Industrial & Systems Engineering of the Georgia Institute of Technology.

Lingzhou Xue, PhD, is Professor of Statistics at The Pennsylvania State University.

Jasen Zhang is a doctoral student in biostatistics at the University of California, Los Angeles.

Introduction

Gun Violence: The Statistical Issues

Gun violence is a vexing problem in the United States. Each year, tens of thousands of Americans are killed by guns, and an even larger number suffer life-altering injuries. Despite this incredible toll, many basic facts about gun violence remain unknown. Scholars are actively seeking to better understand aggregate levels of firearm ownership, the causes and consequences of firearm use, as well as the effectiveness of longstanding public safety interventions. This situation leaves policymakers with an incomplete working picture, complicating already difficult policy discussions around regulatory restrictions on access to firearms. As an illustration of this reality, a recent review of policy research on the effectiveness of a variety of gun safety regulations by the RAND Corporation found that the available evidence on most policies was either limited or inconclusive (Smart, 2024). This state of affairs suggests the need for greater scholarly attention to be focused on this subject.

The present volume emerged from a National Institute of Statistical Sciences (NISS) Ingram Olkin Forum (IOF), held on May 10, 2024, which itself was a follow-up to the first IOF on the topic held on June 26, 2019, in Alexandria, VA. These forums and the resulting papers explore the important ways in which statistics and statistical issues contribute to the study of gun violence, especially in the United States. While many disciplines add to our knowledge of gun violence, statisticians and other quantitative social scientists contribute two critical pieces of knowledge infrastructure – data and models – that are essential for making sense of how much gun violence occurs in our society and how effective our policies are at reducing it.

Gun violence data comes in many forms – vital statistics collected by the public health system, crime reports produced by police departments, as well as newer open-source and crowd-sourced databases maintained by news and other non-governmental organizations. Oftentimes, these multiple sources of information provide a common picture, but sometimes they do not.

Compiling, assessing, and comparing these data sources form the bedrock analytical work of statistics. This book provides an overview of existing gun violence data in use by statisticians today, its strengths, and its limitations.

The other core analytical challenge in gun violence research is correctly modeling and inferring policy effects, given the incompleteness of available

DOI: 10.1201/9781003430773-1

data and the absence of randomized control trials to aid in the assessment of interventions. This book includes several cutting-edge methodological studies that illustrate the emerging possibilities for rigorously making sense of this data and the effects of these policies.

In addition to describing these recent statistical efforts, the book also aims to engage new statisticians, data scientists, and other social scientists to work across disciplinary lines to help understand and mitigate the effects of gun violence. Compared to many other policy areas, the barriers to entry for interested researchers are surprisingly low. Most, if not all, of the relevant data sources are publicly available and, when analyzed carefully, can yield important policy-relevant results with a minimum of effort. And as the chapters in this volume make clear, this work is vital, challenging, and a work in progress.

Chapter 1 – While fatal shootings have long been captured by both vital statistics published by the Centers for Disease Control (CDC) and crime statistics published by the Federal Bureau of Investigation (FBI), non-fatal shootings have been largely missing from most statistical databases despite their significantly greater frequency. The first chapter of the volume, written by Catherine Barber, Phil Cook, and Susan Parker, describes ongoing efforts to improve the availability of non-fatal shooting data at the state and federal level. These attempts rely largely on emergency room and other health sector data rather than more traditional law enforcement data sources. The chapter highlights the opportunity to repurpose existing data series to provide greater visibility into a historical "dark figure," a quantity we have long known that we do not measure very well.

Chapter 2 – In the second chapter, Cook provides a sobering study of the odds of case solving in fatal and non-fatal shooting cases in Raleigh, Durham. Despite similar origins, fatal and non-fatal shooting cases are often treated very differently by law enforcement agencies. In most jurisdictions, the major efforts to solve shooting incidents focus on cases resulting in a fatality, whereas any shooting incident, whether resulting in death or not, can provide information about motive and the decision to use a gun to commit a crime. This work highlights the value of examining the entire distribution of shooting cases rather than the much more limited subset of these cases that result in a fatality. It also shows that most shootings go unsolved and unpunished, thereby undermining all but the most optimistic hopes of preventing further shootings via deterrence or incapacitation.

Chapter 3 – The third chapter explores another "dark figure" of underreported gun-related incidents: shootings, near-misses, and other gun discharges that are only incompletely reported to police. Acoustic gunshot locator systems (AGLS), also known as gunshot detection technology (GDT), has recently become more widely deployed within US cities, particularly those facing longstanding gun violence problems. This technology, which promises to assist law enforcement in identifying the location of gunfire incidents occurring within covered areas, has opened a new window into the spatiotemporal distribution of unreported or incompletely reported gunfire incidents. It has also raised a number of important evaluation questions. How accurate is this new method of locating gun incidents? And what has been its impact on law enforcement effectiveness and gun violence when it is used? Lawrence and Piza review what is known about how these sensor-based systems for detecting and locating firearm discharges have impacted the ways in which police departments respond to and resolve gunfire incidents. Importantly, given the somewhat mixed findings in the existing literature, they highlight the ongoing research opportunities in this expanding knowledge area.

Chapter 4 – In response to several recent high-profile police-involved shootings, where civilians were injured or killed by police gunfire, a sizable literature has now revisited the causes, correlates, and trends in these events. The authors of this chapter, Nix, Adams, and McCrain, examine the state of this literature. In particular, they explore the benefits and trade-offs between classic and emerging sources of data on police-involved shootings, the sensitivity of certain questions to the sources of data that are used to answer them, and the benefits of using formal methods, including direct acyclic graphs, to guard against the risks that measurement error will bias resulting estimates. In making these points, they reinforce the core challenge of measurement error in gun violence research. They also illustrate the ways in which newer sources of data, many of them crowd-sourced in response to the limitations of existing official statistics, have their own measurement limitations, which are not always as well understood. Ultimately, validation and cross-validation will be needed to ensure that police-involved shootings, as well as other sources of gun violence data, are valid and useful for all analytical purposes.

Chapter 5 – Mentch and Cecil build on Chapter 4 and show the need for care around the very issues concerning police shootings that Nix et al. outline conceptually. They show that conventional estimators of racial disparities in police-involved shootings are highly sensitive to the denominator used to estimate these cross-race differences. Relying on US Census estimates of the population at risk, which in turn is based largely on the resident population of areas, leads to large estimated disparities. Substituting more conservative estimates of the population at risk, which often involves using justice-involved population estimates of the population at risk, leads to much smaller estimated disparities. Inattention to this nuance has led to some confusion in the literature, mirroring parallel "denominator" disagreements in other racial disparities studies related to police-initiated traffic stops.

Chapter 6 – Loeffler explores whether ammunition from firearms recovered by police after making arrests ranging from traffic stops to armed robberies, a previously unused source of data, could be systemically examined to infer features about the market for illegal ammunition. His analysis shows that indeed this sort of information can be used to test at least some hypothesized features of illegal ammunition markets. His analysis also shows that novel data sources can shed new light on old questions, which is especially important since traditional sources of data on the diversion of legally sold firearms and ammunition into the illegal market for those goods is an incompletely observed and understood process. This chapter also provides some evidence of the rarity of a posited behavior for carrying an illegal firearm – namely, carrying unloaded firearms for the sole purpose of defensive brandishing or coercive threatening.

Chapter 7 – Ben-Michael et al. show how recent developments in causal estimation are especially relevant to gun violence research. While multistate panel studies have long been used to estimate the impact of specific gun policies on crime, they show that new causal effect estimation tools, such as augmented synthetic control methods, staggered difference-in-difference models, as well as policy emulation frameworks allow for much more robust evaluation of gun policies across different states. Using these methods they illustrate how much more reliably the effect of right-to-carry laws on violent crime can be estimated using these methods than is the case when traditional panel data methods, such as difference-in-difference models, are used.

Chapter 8 – Ko, Zhang, and Holbrook review the framework of Hawkes processes, a class of self-exciting stochastic point processes widely applied to phenomena characterized by contagion. They highlight the computational challenges that arise in fitting these models and then demonstrate the application of a recently developed high-performance computing-powered Bayesian inference strategy to over 400,000 incidents of US gun violence (2014–2024). This reinforces how something as conceptually simple as modeling the interdependence of shooting events in space and time can benefit from innovative improvements in statistical computing.

Chapter 9 – Illustrating why the developments in statistical computing, such as those covered in Chapter 8, are so valuable, Dong and Xie develop a non-stationary spatial-temporal point process model to extend traditional Hawkes processes by introducing a neural network–based non-stationary spatial kernel and incorporating socioeconomic covariates to capture community-level heterogeneity and demonstrate its predictive performance and interpretable spatial contagion effects in an empirical study of nearly 16,000 gunshot incidents recorded in Atlanta between 2021 and 2023, thereby demonstrating the surprising amount of predictability of social processes that are often considered quasi-random.

Chapter 10 – Oliver et al. also deploy the Hawkes contagion model to explain the increased rate of mass shootings in the US since the start of the COVID pandemic. Although some mechanisms are not well understood, they explore how the reproduction number, or contagiousness, of mass shootings has increased since March 2020. They show that the reproduction rate of mass shootings increased even after controlling for the overall levels and a variety of time-varying covariates. In keeping with arguments offered in Chapter 4, they show that their results are robust across a range of data sources. They also posit that aspects of human activity patterns, disrupted by the early years of the pandemic, likely offer the best explanation for the observed change in the reproduction number.

Across all of these chapters, a common theme emerges. While the scale of the gun violence problem is large and the data we have at our disposal is incomplete, much can be learned about gun violence and effective strategies for reducing it (as well as ineffective strategies for reducing it) if the available data is carefully analyzed with attention paid to the known limitations and quality of the available data and using the best contemporary models for analyzing it.

While public attention can turn from one form of gun violence, for example, drive-by shootings, to another form, such as mass shootings, the core analytical problems across all of these sub-components remain shared. Fatal shootings are a random subset of all shootings, and near-miss/attempted shootings likely dwarf both fatal and non-fatal categories. Thus, how data are collected and analyzed is crucial for understanding gun violence. The number of shootings in any given year can fluctuate and is driven by many factors related to human activity patterns, access to firearms, and the effectiveness of public policies designed to restrict prohibited users from accessing firearms.

As with other areas of policy science, the area of gun violence research has experienced repeated cycles of high-profile findings that have not turned out to hold up as well as initial attention suggested they might. Early results on concealed carry laws and police involved shootings, both sub-literatures beginning with strong statements in support of unexpectedly large results. Subsequent works, including work in this volume, have shown that these results are not robust. Small tweaks to data sources, model specifications, or models all lead to generally null results, indicating that single studies or even multiple studies, when weighed against many studies, rarely make for robust results. Recognizing this reality highlights the need for more high-quality published studies, so that the next time the RAND Corporation, or another author, publishes a systematic review on gun policy research, the expanded literature can enable a more definitive conclusion on each of the many policies that we employ.

References

Smart, R., et al. (2024). *The Science of Gun Policy: A Critical Synthesis of Research Evidence on the Effects of Gun Policies in the United States*, Fourth Edition. Santa Monica: The Rand Corporation. https://www.rand.org/pubs/research_reports/RRA243-9.html

1

Nonfatal Firearm Injury Surveillance in the US: An Update

Catherine Barber, Philip J. Cook, and Susan T. Parker

1.1 Introduction

The harm done by firearms extends far beyond the nearly 50,000 people killed by gunfire each year. In 2022 alone, over 98,000 nonfatal firearm-related injuries were treated in emergency departments (Table 1.1). The 2024 US Surgeon General's Advisory statement calls gun injury a public health crisis that requires the nation's immediate awareness and action (Office of the US Surgeon General, 2024). Over half of US adults report that they worry about a loved one becoming a victim of gun violence (Schumacher et al., 2023). Firearm injuries have emerged as the leading mechanism of death among young people in the US (Goldstick et al., 2022).

Despite the substantial burden of harm from nonfatal shootings, no single source of administrative data tracks all nonfatal shootings in a systematic, reliable, and timely fashion (Cook & Roman, 2021). Data on nonfatal firearm injuries are tracked in two primary sources: hospital emergency department records and law enforcement records. Both have intrinsic limitations. Hospital data are usually limited to cases in which the victim received treatment, while law enforcement data are often limited to criminal assaults, omitting accidents and self-inflicted shootings. In practice, there are also deficiencies in the information included in the most prominent national compilations: hospital data are unreliable with respect to the intent of the shooting (M. Miller et al., 2022), while the Federal Bureau of Investigation (FBI) compilation of law-enforcement data does not distinguish between gun-assault cases in which the victim is shot and cases in which the victim is pistol whipped or otherwise injured. Modest reforms could do much to improve both of these sources (Barber et al., 2022; NORC, 2020).

 DOI: 10.1201/9781003430773-2

TABLE 1.1

Deaths and Estimated Hospital-Treated Injuries from Firearms in 2022, United States

Intent	Deaths	Case Fatality Rate	Injuries (Approximate)
Unintentional	463	0.06	7,254
Assault	19,651	0.19	83,775
Legal Intervention	643	0.55	526
Suicide/self-harm	27,032	0.85	4,770
Undetermined	415	0.15	2,352
Total	48,204		98,677

Notes: We use 2022 firearm death data from the CDC's WISQARS fatal data reports. We apply case fatality rate estimates from the literature to the count of deaths to estimate the approximate number of nonfatal firearm injuries.

The current systems for recording deaths from firearm violence provide detailed, comprehensive, and reasonably accurate data. Some mortality measures are available to the public within a matter of days, such as homicide counts gathered by local police agencies. But mortality data, while essential, are not enough for the purposes of public health surveillance and evaluation, since most gunshot victims do not die. The aspirational goal for improved data systems for nonfatal gunshot injury is to approximate the quality of current systems for recording fatalities.

A complete nonfatal firearm injury surveillance system should accurately record instances of gunshot wounds for each type of intent, including accident, assaultive, self-harm, police-involved/legal intervention shootings, and undetermined. To maximize its usefulness, this data should be compiled and made available to the public on a timely basis with appropriate safeguards for confidentiality.

A comprehensive firearm injury surveillance system has several important functions:

- Early warning systems – An ideal firearm injury surveillance system would at the most basic level be able to provide information on whether gun injuries are currently increasing or decreasing at the neighborhood level and by victim demographics. Nonfatal firearm injuries, which occur twice as often as fatalities, are particularly valuable for detecting meaningful shifts in trend in areas where gun injuries are less common.
- Epidemiological descriptions of nonfatal firearm injury – Detailed firearm injury surveillance data would expand the contextual circumstances of nonfatal gun violence victims, particularly as it concerns information unavailable in an emergency department but

known to law enforcement or social service providers. Detailed epidemiological descriptions of nonfatal firearm injury could further enhance what is known about the determinants of fatal v. nonfatal firearm injuries and could help determine, for example, whether a change in the number of firearm homicides is driven by a change in overall assaultive shootings or in the case of fatality rate for those shootings.

- Causal inference and program evaluation – Combining data on both fatal and nonfatal firearm injury provides researchers far greater statistical power to evaluate what works to reduce gun violence at the neighborhood level than using death data alone.

Updates are underway to improve both hospital and law enforcement data sources to better address surveillance needs and take steps to address some of these important functions. However, substantial challenges remain to achieving comprehensive national nonfatal firearm injury data, and these challenges may mean substantial delays in access for communities, researchers, and other stakeholders as they seek to prevent and reduce firearm injury. This chapter reviews the case for measuring nonfatal firearm injuries, surveys literature measuring nonfatal shootings, and examines how firearm data infrastructure may enable or impede advances in what is known about gun violence.

1.2 Annual Incidence and Intent of Nonfatal Shootings

In 2022, 48,830 Americans lost their lives to gun-related injuries, according to annual mortality data files that the Centers for Disease Control and Prevention (CDC) maintains (WISQARS Fatal and Nonfatal Injury Reports, 2022). Systems that gather fatal firearm injury data are in place and reliable in total count. Some inaccuracy in classifying intent exists in the National Vital Statistics System data (which is the nation's official mortality data and is based on death certificate information), especially with respect to police shootings and unintentional firearm injury deaths. The federal source that accurately classifies fatal firearm injuries is the National Violent Death Reporting System (NVDRS) (Barber & Hemenway, 2011; Conner, Azrael, Lyons, et al., 2019). While NVDRS now provides data for all 50 states, a few states do not cover all counties in their state, and there is typically a two-year lag before data are released.

Because no current system comprehensively tracks nonfatal injuries by intent, we estimate nonfatal injuries in 2022 by dividing deaths by intent-specific "case fatality rates" drawn from the research literature (Table 1.1)

(Barber et al., 2022). We apply a case fatality rate of 19% for firearm assault (Braga & Cook, 2018; Kaufman et al., 2019), 55% for legal intervention (police shootings) (Nix & Shjarback, 2021; Ward et al., 2024), 85% for suicide (Conner, Azrael, & Miller, 2019), 6% for unintentional shootings, and 15% for undetermined (Barber et al., 1998; Vyrostek, et al., 2001). Of fatalities in 2022, the majority of firearm deaths are suicides (56%), followed by assaults/legal intervention (42%), unintentional (<1%), and undetermined (<1%). For nonfatal injuries, however, the majority of nonfatal shootings are assaultive in nature, followed by unintentional injuries, suicide, and undetermined.

1.3 Why Measure Nonfatal Firearm Injuries?

Measuring nonfatal shootings impacts our understanding of the burden of harm from firearms. Whereas most fatal shootings are suicides, the majority of nonfatal shootings are assaults. This is due to the lethality of suicide attempts with firearms, as 85% of attempts are fatal. In contrast, victims survive in approximately 85% of firearm assaults, 95% of unintentional shootings, and 45% of legal intervention (police shootings). In this sense, existing fatal injury surveillance may be adequate for surveillance of firearm suicide, but not for other intents.

Nonfatal shootings data could expand recent single-city analysis of increasing case fatality rates in Chicago (Ludwig & Miller, 2024) and Los Angeles (Brantingham et al., 2024), providing valuable evidence, as Ludwig and Miller suggest, concerning high-capacity magazines and devices to turn semi-automatic firearms into automatic weapons. More broadly, detailed data on nonfatal shootings could expand what is known about case fatality rates, making it possible to understand when they are impacted by medical advances, changes in firearm or ammunition design, firearm policies, or public safety initiatives.

Comprehensive firearm injury surveillance data would also enhance efforts to evaluate programs, policies and laws aimed at reducing gun violence. Currently, studies often rely on data obtained from law enforcement. Not all shootings are reported to police, and law enforcement definitions are not always comparable across jurisdictions. In the same way nonfatal firearm injury data would enhance the capacity of early warning systems to detect changes and patterns in gun violence, policy evaluation efforts would be better powered to detect statistically significant effects of gun violence reduction efforts.

As Lauritsen and Lentz (2019) note in their comparison of several decades of fatal and nonfatal violent crime sources, variation in homicide rates may result from several factors, each with different policy implications. Homicide rates could rise due to an overall growth in violent crime, an increase in gun availability leading to a greater proportion of crime committed with a firearm, or an increase in the case fatality rate for gunshot injuries. Increases in case fatality, in turn, can result from change in availability and quality of medical care, changes in gun and ammunition type, changes in specific sub-type of assault (e.g., domestic violence shootings typically have higher case fatality rates than street shootings), etc. It is important for policy to differentiate between these pathways as each suggests a different strategy for violence reduction and none are possible to examine without nonfatal data.

Similarly, changes in overall suicide rates cannot be interpreted without data on nonfatal suicide attempts. An increase in suicide rates may result from an increase in suicidal distress (nonfatal injuries by all methods climb in tandem with fatalities), a decrease in access to effective medical care (nonfatal injuries remain flat while fatalities climb), or an increase in at-risk people having access to firearms. As Barber and Miller (2014) have noted, given low case fatality rates of suicide attempts with pills or cutting, even a small proportion of at-risk individuals moving from those methods to higher-lethality methods (especially firearms) can dramatically increase the suicide rate.

Finally, tracking nonfatal firearm injuries is an essential first step in identifying victims who could be served by in-hospital programs that aim both to help victims get the physical and mental health care they need and to help them avoid both revictimization and retaliation. Most hospitals will not discharge a patient who has attempted suicide without first providing a mental health evaluation and offering treatment options. But until very recently, few hospitals offered interventions to victims of violent assaults. Emerging literature suggests that in the aftermath of a gunshot wound injury, young victims and their families have increased need for mental health services and families incur additional medical costs (Lei et al., 2023; Magee et al., 2022, 2023; Song et al., 2023). Examining nonfatal shootings by intent informs both prevention efforts and resource allocation to best support survivors.

1.3.1 Inaccurate Measures of Nonfatal Firearm Injuries

A lack of data on nonfatal firearm injuries has led to inconsistencies and even inaccuracies in the public health, criminology, and medical literatures.

One source of problematic nonfatal firearm injury data comes from hospital-based sources. These sources are prone to inaccurate recording of firearm injury by intent (Barber et al., 2022; Hunt et al., 2007; MacPhaul et al., 2023). Hospital data sources that use the International Classification of Diseases (ICD) coding systems often record firearm injuries that are in fact assaults as accidental or unintentional injuries. One result is to overstate the number of accidental shootings and underreport assaults. Scholars in the medical literature, relying on this misleading data, have gone on to incorrectly state that accidental injuries are the leading cause of firearm injury in the US (Kaufman et al., 2021, 2024)

At times, problems with nonfatal firearm injury data have garnered public and congressional attention. Kalesan et al. (2017) reported a trend of increasing case fatality rates for firearm injuries from 2001 to 2013, a period during which firearm homicides were stable, based on nonfatal firearm injury data from the Centers for Disease Control's National Electronic Injury Surveillance System – Firearm Injury Surveillance System (NEISS-FISS). However, after adjusting for several problematic features of the NEISS-FISS sample, Cook et al. (2017) found that firearm injury rates remained constant during the period. NEISS-FISS estimates are generated from a small survey of US emergency departments, and estimates are now denoted as "unstable" on the CDC's public query access portal.

Inconsistencies in nonfatal firearm injury measures also arise in the medical literature and can stem from the type of hospital setting used in the analysis. Different medical settings may be used to treat different types of firearm injuries. Patients with more severe firearm injuries may be routed to trauma centers better suited to treating life-threatening medical emergencies. Alternatively, in more rural areas where no trauma center may be readily accessible, an emergency department may treat severe firearm injuries. Because different places of service may treat different injury severity, estimates of case fatality rates may be affected by whether a study samples hospitals, trauma centers, or urban vs. rural settings. As such, estimates from trauma centers of case fatality rates for assaults range from 11% (Tessler et al., 2019) to inpatient settings at 8% (Agarwal, 2015). Because hospital-based case fatality estimates are only able to consider patients who live long enough to be treated in a hospital setting, thus omitting victims who die at the scene of a shooting, they will always tend to fall below the overall case fatality rate for assault of approximately 20%.

1.4 Data Sources for Tracking Nonfatal Firearm Injuries

Data on nonfatal firearm injuries come from two sources: criminal offenses reported to police and emergency department visits for gunshot wounds. Hospital data is promising and is standardized in several useful ways. However, hospital data contains a number of inaccuracies and are not timely for surveillance. Police data has the potential to track gun assault injuries, and a number of police departments do report the number of nonfatal criminal shootings.

Other sources of data on nonfatal shootings exist, but they are not considered in this chapter as they are not reliable or comprehensive. In particular, researchers increasingly rely on the Gun Violence Archive (GVA) which gathers data on fatal and nonfatal shootings from public police reports and media sources. While the GVA gathers fairly reliable data on fatal firearm injuries, the same is not the case for nonfatal firearm injuries. In part, this is due to fluctuations in media coverage of nonfatal shootings, which can vary substantially over time and suggest misleading trends in nonfatal gun violence within a jurisdiction (Magee et al., 2024). Underreporting in the GVA is systematic and varies by victim and offense attributes (Gobaud et al., 2023; Magee et al., 2024). It is likely that reporting also differs by jurisdiction as some jurisdictions will have more comprehensive press coverage. As such the GVA is not considered in this chapter.

The National Crime Victimization Survey (NCVS) is also not considered. Underreporting of nonfatal firearm injuries incurred in criminal assaults has persisted since it was first documented nearly four decades ago (Cook, 1985). In 2021, only 23,007 (+/−16,685) victims of nonfatal firearm assaults were estimated in the national sample (*NCVS Dashboard: Multi-Year Trends Crime Type*, n.d.). While this figure more closely tracked reality in 2022 with an estimate of 78,876, variation in year-to-year estimates renders them unusable for tracking nonfatal firearm injuries.

1.4.1 Hospital Data Sources

Hospitals send data to statewide hospital inpatient and emergency department databases, and nearly all states, in turn, submit these databases to the national repository of health data run by the Agency for Healthcare Research and Quality (AHRQ). This robust data infrastructure is standardized because it uses uniform hospital billing system variables and code values. Diagnosis codes are established by the International Classification of Disease, currently in its 10th revision (ICD-10-CM). For injury visits, a billing coder assigns ICD diagnosis codes that describe the injury type (e.g. laceration, fracture) and body part affected, as well as an "external cause" code (E-code) that describes both the mechanism of injury (e.g., firearm, sharp instrument) and intent (e.g., assault, accident/unintentional, suicide/self-harm, legal intervention, undetermined).

Hospital data is likely to record a substantial proportion of gunshot wounds as patients are likely to seek treatment if injured with a firearm. Survey data of jail inmates suggests that over 90% of those who had been shot reported seeking hospital treatment for their gunshot injuries (May et al., 2002). Some victims may not seek professional care, and a small minority of gunshot wounds, approximately 4%, may be treated in urgent care settings, which will not be captured in any hospital record (Gilbert et al., 2024). Additionally, hospital billing systems may not capture all instances of firearm injury in E-codes as this designation is not necessary for reimbursement, though evaluations suggest that over 85% of injuries that should have an E-code do have one (Barret, 2016; Stewart et al., 2024).

It is possible to combine estimates to obtain an approximate calculation of the proportion of total gunshot wounds treated in hospitals. If approximately

90% of individuals suffering gunshot wounds seek hospital care, of which 4% seek non-ED care, and 85% of the gunshot wounds are correctly counted, then gunshot wounds recorded in hospital billing data represent approximately three quarters of the true total of all gunshot wounds.

1.4.1.1 The National Emergency Department Sample (NEDS) and State Emergency Department Databases (SEDD)

The National Emergency Department Sample (NEDS) consists of a large sample of US emergency departments and can provide relatively precise national and regional estimates of how many emergency department visits occur for nonfatal firearm injuries. Gathered by the Agency for Healthcare Research and Quality (AHRQ) Healthcare Cost and Utilization Project (HCUP), these data are freely available via the HCUPnet (https://datatools.ahrq.gov/hcupnet/) reporting tool, AHRQ's portal for accessing summary estimates and counts of its data holdings. Descriptive data on victim demographics including age and sex (though not race) are made available via HCUPNet. However, there is a 2–3 year delay in posting these data. Researchers who have taken HCUP's data use training can purchase de-identified, individual-level NEDS data.

While NEDS provides broad estimates of national nonfatal firearm injuries, the State Emergency Department Databases (SEDD) and State Inpatient Databases (SID) record information for all visits to emergency departments and inpatient settings in each state that participates. SEDD records all emergency department visits that are not transferred to an inpatient setting, while those visits that are admitted to an inpatient setting are recorded in the SID, necessitating combining the two for a comprehensive count of firearm injury. Nearly all states now report comprehensive data in the form of SID and SEDD to AHRQ annually. For surveillance of state-level gunshot wounds, annual count data with demographics of firearm victims are available through HCUPNet. For more detailed data that would include diagnosis, sub-state geography, or more granular time of injury, the prohibitive costs of access to the SEDD and SID are a barrier. Access to all 26 states that have submitted data for 2020 would cost a non-profit purchaser $24,348 dollars for a single year of SEDD access.

Efforts are underway to improve how injury intent is coded in hospital data records. Annually, the Centers for Medicare and Medicaid Services (CMS) and the National Center for Health Statistics (NCHS) in the Department of Health and Human Services (DHHS) provide updated guidelines for ICD coding in healthcare facilities. For years the guideline has been that when injury intent is not specifically indicated, coders should default to "accident," which is likely to be correct for most injury types. However, for nonfatal firearm injuries, assaults are the predominant type. An amendment to the guidelines for medical facilities was proposed to change the default from *accidental* to either *assault* or to *undetermined intent* (ICD-10-CM Official Guidelines for

Coding and Reporting, 2023). NCHS/CMS opted to change the default for firearm injuries to undetermined intent effective October 1, 2025.

This recent change is likely to continue the undercount of assaults in hospital data. The American Hospital Association, which issues specific coding guidance to hospital billing coders, will undertake an effort to educate coders about key words and phrases that can trigger coding cases as assault and that differentiate assaults, accidents, and self-harm. This education may solve some of the coding problems. But given evidence that coders and coding software sometimes automatically apply defaults to E-codes even in the presence of evidence of intent in the record, this education may be insufficient (Barber, 2021; M. Miller et al., 2022).

1.4.1.2 National Electronic Injury Surveillance System (NEISS) Firearm Injury Surveillance System (FISS)

Since 1971, the Consumer Product Safety Commission (CPSC) has maintained a survey of approximately 100 emergency departments with abstractors in each emergency department tasked with recording detailed data on product-related injuries. In 1993, the CDC partnered with the CPSC to expand the data collection to product-related injuries that include firearms (*National Electronic Injury Surveillance System – Firearm Injury Surveillance Study (NEISS-FISS) – Healthy People 2030 | Health.Gov*, 2024).

Unlike HCUP data sources, NEISS-FISS does not obtain injury intent coding from hospital billing data; instead, abstractors contracted by NEISS review patient records and enter coded data summarizing the encounter and write a sentence describing how the injury occurred (NEISS Coding Manual, 2019). A small group of coders at the CDC injury center reviews narratives for accuracy of mechanism and intent coding. As such the proportion of injuries by intent for firearms is more accurate than the billing coding process on which NEDS and SEDDS are based. NEISS-FISS gathers detailed information spanning traditional medical data fields as well as fields beyond what is normally in claims data records, including the location where injury occurred and a brief narrative describing the incident.

The sample of NEISS emergency departments from which its public-facing data query website draws has historically produced unstable national estimates of firearm injuries (*WISQARS Nonfatal Injury Reports*, n.d.). The reasons include small sample size (especially before 2023) and, more importantly, extreme clustering of assault-related gunshot wounds by hospital because approximately 6% of hospitals account for two-thirds of all gunshot wound cases (Barber, 2021), leading to confidence intervals far too broad to generate stable national estimates. Additionally, as discussed in detail in Cook et al. (2017), changes in NEISS abstractors practices have resulted in lower proportions of unknown injury intent in recent years. Second, because the sample of hospitals is not stable over time, hospital substitutions have resulted in large swings in estimates of firearm injury. In an effort to produce more reliable national estimates, the CDC has focused on increasing the number of hospitals participating in NEISS-FISS and ensuring that hospitals that drop out are replaced by those treating relatively comparable numbers of firearm victims; however CDC currently advises using HCUP's NEDS for surveillance estimates (*WISQARS Nonfatal Injury Reports*, n.d.).

1.4.1.3 National Syndromic Surveillance System (NSSP) and FASTER/AVERT

The National Syndromic Surveillance System (NSSP) gathers data in near real time from approximately 70% of acute care hospitals nationally on the demographics of patients visiting the emergency department and the nature of the visit. NSSP is a promising source of timely data on nonfatal firearm injury. NSSP captured 425 million ED visits, of which 338,390 were firearm-related, from January 2018 to December 2023 (Harrel et al., 2024). However, access to individual-level NSSP data is limited to a set of permissioned public health officials. However, for selected conditions, state officials have worked to make state-level public data available through data dashboards and other means.

In 2020, the CDC provided funding to ten states to pilot test collecting firearm injury data through NSSP, an initiative called FASTER (Firearm Injury Surveillance Through Emergency Rooms). Of the ten states funded, Georgia, North Carolina, Oregon, Virginia, and Washington DC created dashboards to disseminate timely data on firearm injury trends (*DC Firearm Injury Surveillance Dashboard| DC_FASTER*, 2023). Initial findings show that while it is feasible to identify firearm injuries, the information from which to classify intent is often not available or inaccurate.

In 2023, FASTER transitioned to AVERT (Advancing Violence Epidemiology in Real-Time). AVERT collects near real-time emergency department data on firearm injuries, other violence-related injuries, and mental health conditions in 12 states for 5 years (CDC, 2024).

1.4.2 Law Enforcement Data Sources

Police investigate criminal assaults and shootings and record detailed information about the victims, circumstances, and perpetrators. Unlike hospital data, the universe of police data is limited to criminal assault and does not span all injury intents. While this limits the sample, in practice, assault injuries are the majority of nonfatal firearm injuries. Further, police records give details about the nature and circumstances of an assaultive nonfatal shooting. Unlike hospital data, police agencies gather information on the shooter, the firearm, and the incident location as well as other relevant data beyond the scope of the medical system to track.

It is not clear what proportion of assaultive gunshot wounds are known or reported to police. In theory, the majority of states (45 as of 2014) have enacted mandatory reporting laws that require health providers to report any firearm injuries they treat to law enforcement (Houry et al., 2002). The extent to which providers comply with mandatory reporting is unclear, and research comparing police and hospital-reported gunshot wounds suggest that some wounds are known to police but not to hospitals (Magee et al., 2021; K. R. Miller et al., 2022); others suggest that gunshot wounds are known to hospitals and not police (Kellermann et al., 1996).

Police agencies voluntarily report data to the Federal Bureau of Investigation (FBI). For decades, FBI data was limited to summary offenses and arrests reported to police, which were collected by the FBI in the Summary Reporting System (SRS). The SRS captured robbery and assault counts by weapon type, but did not distinguish between gun assaults or robberies in which a victim was shot and those in which a victim was threatened or otherwise injured with a gun. Therefore, aggregated police data, despite widespread use, did not provide a direct measure of nonfatal firearm injuries. Further, definitions of what kinds of assaultive offenses constitute an aggravated assault with a firearm may differ across jurisdictions (Hipple, 2022).

The SRS system was sunsetted in 2021 and replaced by the National Incident-Based Reporting System (NIBRS). At the time of the transition, nearly half of the nation's law enforcement agencies were unable to report data to NIBRS. While coverage has since expanded to approximately 70% of the nation's law enforcement agencies, smaller rural agencies remain unable to report data to NIBRS (Asher, 2023). In the 2022 filing, 14,631 law enforcement agencies spanning 91.7% of the US population reported data to NIBRS (*FBI Releases 2022 Crime in the Nation Statistics*, 2022). Delays in data processing mean complete crime data is unavailable for nearly ten months after a year has ended.

It is important to note that police agencies could submit comprehensive nonfatal firearm injury data in instances of police shootings to the FBI and fill an important data gap with information most agencies already routinely gather. However, the participation rate among police agencies in the FBI's existing Use of Force repository is so low that the threshold for public dissemination, which does not exist with NIBRS, is not met and data is therefore not released (Federal Bureau of Investigation, 2024).

1.4.2.1 National Incident-Based Reporting System (NIBRS)

NIBRS gathers detailed data from law enforcement agencies about the offenses, arrests, weapons, and offenders involved in the commission of crime in the US. In contrast to the SRS, NIBRS details each crime incident separately, and includes a wider array of crime types. NIBRS expands reporting of firearms associated with additional crimes, adding firearm sexual assault, rape, kidnapping, and extortion. NIBRS may improve the accuracy of recording aggravated assaults committed with a firearm. In a study comparing UCR to NIBRS reporting of firearm crimes in a sample of five states that consistently reported data to NIBRS, 18.8% more aggravated assaults with a firearm were recorded and 2.1% more robberies with a firearm were recorded in NIBRS relative to the summary reporting system (Parker, 2022).

Not only does NIBRS gather detailed, incident-level information, but recent changes to NIBRS may substantially improve the quality of the data that NIBRS gathers about victims of gun violence. In March 2023, the FBI Criminal Justice Information Services (CJIS) Advisory Policy Board (APB) approved two proposed changes to how NIBRS measures gun violence victimization (FBI Criminal Justice Information Services (CJIS), 2022). First, the new injury coding adds the category of "gunshot wound" to the injury type. Second, a new field will track whether a firearm was discharged during the crime.

While these changes have been approved, it is unclear when they will be fully implemented into the technical specifications for data NIBRS gathers. The gunshot wound category is present in the 2025 NIBRS Technical Specification; however, the addition of a firearm discharge category is not. Further, the changes that the FBI APB approved come alongside a myriad of other changes to NIBRS data. With the time needed to train law enforcement officers to record new forms of data, to update records management (RMS) systems to accept the new data, and for state crime reporting agencies to communicate the changes to local agencies, it is unlikely that the improvements to NIBRS will soon be reality for all agencies. A more realistic time frame may be on the order of 5–10 years.

The FBI is not the sole aggregator of local police crime data. Many law enforcement agencies are already reporting crime data in near real time on their own public portals. Portals allow residents to understand granular neighborhood trends in crime, and many agencies already track data on nonfatal shootings. This data is often posted shortly after a crime occurs and can be useful in understanding timely, granular trends in gun violence. Several efforts are underway to aggregate this data for public and research consumption. One example is the NORC Live Crime Tracker, a data dashboard that gathers crime statistics reported by more than 50 of the nation's largest cities on their open data portals (*Live Crime Tracker | NORC at the University of Chicago*, n.d.). Another effort is conducted by AH Analytics, which developed the Real-Time Crime Index (RTCI) that aggregates monthly crime data from 500 to 1,000 local police departments to visualize nationwide crime trends ("Real-Time Crime Index," n.d.).

Finally, state agencies may have an additional role to play in making crime data more available. Most states have Uniform Crime Reporting (UCR) programs that serve as the intermediary between an individual local law enforcement agency and the FBI. State UCR agencies in practice gather and report state-level data to the FBI, rather than agencies reporting directly. Some state UCR programs make their data publicly available, and some aim to do so with a shorter lag in reporting time. Minnesota recently launched a crime data explorer featuring updates from all agencies every month

(Fox-Sowell, 2023). Recently, the state of Illinois passed a bill that aims to increase the rate of crime reporting, an effort that other states may similarly undertake (Borica, 2024).

1.5 Implications of Current Nonfatal Firearm Data Infrastructure: What is Possible? What is Not?

Currently, barriers to comprehensive, timely, reliable nonfatal firearm injury surveillance remain in place despite researcher, practitioner, and stakeholder investment and awareness. Many of the improvements detailed in prior sections are substantial undertakings that show promise, but more work is required. For police data sources, while NIBRS has expanded rapidly to new agencies, the current version of NIBRS does not distinguish gunshot wounds from other types of injuries. For hospital data sources, it is not clear that syndromic surveillance will be publicly accessible even if it is scaled beyond the current 12-state sample. For hospital sources based on billing data (NEDS NEDS, SEDDs, and SIDs), the cost of accessing the individual-level data is a major barrier, and intent classification remains a problem. To assess the extent to which these shortcomings impede the goals of surveillance, we discuss the primary functions of nonfatal shooting surveillance data, which include:

- Early warning systems
- Epidemiological descriptions of nonfatal firearm injury
- Causal inference and program evaluation of policies, laws, and other programs to reduce nonfatal shootings

In early warning systems for firearm assaults, it may not be essential to disentangle firearm injury intent, in which case hospital syndromic data, with its rapid turnaround time, is a useful source. Because the rate of firearm accidents changes very slowly over time, and nonfatal firearm suicide attempts are so small in number, any significant volatility in firearm injuries that an early warning system could detect would be driven in large part by assault. Given current progress, early warning systems could rely on less comprehensive but more timely sources including NSSP and local law enforcement data. These sources benefit from a consistent sampling frame, unlike GVA, which captures only a fraction of nonfatal cases.

Another approach may be to apply case fatality rates to mortality counts to provide useful estimates of nonfatal firearm injuries (see Table 1.1). Case fatality rates tend to be consistent over time within intent type, particularly for assaults (P. J. Cook et al., 2017). Consistent case fatality rates are useful in estimating the burden of nonfatal injuries for different jurisdictions and over time for the nation. Less is known about the extent case fatality rates may differ by jurisdiction, however, and it is unlikely that case fatality estimates based on a national constant can capture local variation, making this type of analysis an unsuitable replacement. For jurisdictions where firearm injuries are infrequent, approximations of nonfatal firearm injuries derived from mortality counts are likely to be noisy and therefore not useful for surveillance purposes.

While hospital data in its various forms is promising for some types of early warning systems, they do not lend themselves well to epidemiological descriptions of nonfatal firearm injury where the objective is to understand efforts to solve nonfatal shootings or understand the essential role of the criminal justice system in preventing gun violence (Braga, 2022). Police data are therefore essential to understand the epidemiology of nonfatal gun violence, particularly as it concerns perpetrators and the ability of police investigators to solve crimes. Further, HIPAA compliance means hospitals may be extremely cautious about accessing geographically relevant units of analysis and may not permit relevant linkages to other sources of administrative data. This is unfortunate as smaller areas of analysis benefit the most from nonfatal firearm injury data provided by hospital sources.

An exception to hospital data limitations for epidemiological studies is NEISS-FISS, which provides rich epidemiologic data on firearm injuries, and does so with accurate classification of intent and no cost barriers to researchers. It cannot serve as an early warning system, given its time delay, nor can it provide state- or local-level data. The national estimates it produces have large confidence intervals.

Finally, for program evaluation and causal research purposes, accurate measures of firearm injury inclusive of nonfatal injuries is important for detecting what programs and policies work to reduce gun violence. Accurate measures of firearm injury intent may not be essential for some types of program evaluation and impact studies. For studies that seek to evaluate the effects of child access prevention laws, for instance, one would expect that safer firearm storage could reduce firearm assaults, accidents, and suicide attempts involving young people. In similar instances, access to state-level HCUP State Emergency Department Databases (SEDD) and State Inpatient Databases (SID) is a promising avenue for comprehensive firearm injury data for causal inference and program evaluation as each state now reports the universe of ED visits to HCUP. Taken together, SEDDs and SIDs provide the most comprehensive information at granular geographic units of analysis (e.g., patient residence zip code) for describing the incidence and medical burdens of firearm injuries and the demographic characteristics of victims. The data are not timely, however, and remain prohibitively expensive to purchase in total. NEDS provides access to the same rich data in a more cost-effective way (although not free), but, since it is based on a sample, at the sacrifice of geographic granularity.

Police data will remain essential in understanding the most prevalent form of nonfatal firearm injury: assaults. Police data improvements including recording if a gun is discharged and whether a victim suffers a gunshot wound will better characterize the nature of nonfatal gun assaults. Given the rapid improvements in the adoption of NIBRS, however, there may be reason for optimism about improvements in law enforcement data collection.

Conclusion

As nonfatal firearm data access continues to expand, an important avenue of future research is to understand the extent to which the universe of nonfatal firearm injuries differs by source. It is not known to what extent police and hospitals each comprehensively capture the totality of firearm injuries. In principle, mandatory reporting laws should account for discrepancies in assault counts between hospital and police sources, but it is not known to what extent hospitals follow mandatory reporting laws and police agencies record these reports. It is important to understand the potential for underreporting to law enforcement, whether it varies over time, and whether hospital data mitigates reporting challenges.

Ultimately, monitoring and understanding nonfatal firearm injuries will require detailed data from both police and hospital sources. A complete nonfatal firearm injury surveillance system should accurately and rapidly record instances of gunshot wound. One potential avenue is using NEDS and statewide hospital databases to estimate the total count of firearm injuries. Working to ascertain which firearm injuries are known to police as assaults could address incorrect hospital intent reporting while gathering important details on perpetrator, weapon, and circumstances that only police gather. At the same time, near-real time counts of firearm injuries by victim demographics (but not by intent or circumstances) could be supplied at the local, state, and national levels by the National Syndromic Surveillance System to detect rapid changes. Jurisdictions experiencing high levels of gun violence could similarly provide counts of firearm injuries on a more rapid scale.

An aspirational but achievable goal for nonfatal firearm injury surveillance in the near- to mid-term is to expand the NEISS-FISS system by linking records on firearm injuries treated in its 100 or so emergency departments with police reports to gain information on circumstances, specific weapon type, and, if applicable, perpetrator characteristics and arrest information. While this would generate only national estimates and would not be timely, it would accurately classify intent and collect valuable data on the full range of intent types (Table 1.2).

TABLE 1.2

Summary of Data Source Attributes

Data Source	Sample	Coverage	Accuracy of Intent Coding	Data Rapidly Released	Data Availability	Limitations
NEDS	Nationally representative sample of US state-level SEDD hospital data sources	GSWs treated in ED	Low. Assaults often coded as Unintentional	No – reporting delay	2006–2023. Limited detailed data is available without purchase of NEDS	Outcomes unavailable at units lower than the region
SEDD and SID	State-level samples of all treated ED patients in a state for 36 states	GSWs treated in ED	Low. Assaults often coded as Unintentional	No – reporting delay	1990–2023. Limited detailed data is available; some states have implemented a unique patient ID; many states do not measure race of patient	Intent coding is greatly distorted
NEISS-FISS	100 emergency departments for all firearm injury types	GSWs treated in ED	High. Intent is reviewed by abstractors	No – reporting delay	1993–2023. Detailed case narratives and victim data	Only provides national estimates. Time series is unreliable
FASTER: AVERT	All emergency departments in 12 states	GSWs treated in ED	Low	Yes – supposed to be near real time	Limited patient-level detail	Unrepresentaive sample

(Continued)

TABLE 1.2 (CONTINUED)

Summary of Data Source Attributes

Data Source	Sample	Coverage	Accuracy of Intent Coding	Data Rapidly Released	Data Availability	Limitations
NIBRS	As of 2022-present, population coverage of over 90% of the US	Gun assaults with injury	High	No – 9–11 month reporting delay	1991–2024. Yes – detail on victim, offenders, injury, case clearance	Incomplete coverage until recent years
Local Law Enforcement Agency Data	Single Agency	Gun assaults with (GSW) injury	High	Yes – data is often populated weekly	Varies by agency. Data often has detailed demographics	Reporting differs widely among agencies
NCVS	Nationally representative household sample for victimization reporting	Gun assaults with GSW	High	No – undercounts gunshot wounds and reporting delay	1973–present. Yes – detail on victim reporting, economic and household impact of victimization event	Large underestimate due to non- response and sample issues

Acknowledgment

We thank John Roman for valuable comments on this manuscript. We also thank Charles Loeffler and James Rosenberger for valuable comments.

References

Agarwal, S. (2015). Trends and burden of firearm-related hospitalizations in the United States across 2001–2011. *The American Journal of Medicine, 128*(5), 484–492.e1. https://doi.org/10.1016/j.amjmed.2014.12.008

Asher, J. (2023, August 11). Who reported NIBRS data in 2022? [Substack newsletter]. *Jeff-Alytics*. https://jasher.substack.com/p/who-reported-nibrs-data-in-2022

Barber, C. (2021). Improving the capacity of hospital emergency department data systems to track nonfatal firearm injuries. In John K. Roman & Philip J. Cook (Eds.), *A blueprint for a US firearms data infrastructure: Implementation papers*. NORC at the University of Chicago.

Barber, C., Cook, P. J., & Parker, S. T. (2022). The emerging infrastructure of US firearms injury data. *Preventive Medicine, 165*, 107129. https://doi.org/10.1016/j.ypmed.2022.107129

Barber, C., Goralnick, E., & Miller, M. (2021). The problem with ICD-coded firearm injuries. *JAMA Internal Medicine, 181*(8), 1132–1133. https://doi.org/10.1001/jamainternmed.2021.0382

Barber, C., & Hemenway, D. (2011). Too many or too few unintentional firearm deaths in official U.S. mortality data? *Accident; Analysis and Prevention, 43*(3), 724–731. https://doi.org/10.1016/j.aap.2010.10.018

Barber, C. W., & Miller, M. J. (2014). Reducing a suicidal person's access to lethal means of suicide: A research agenda. *American Journal of Preventive Medicine, 47*(3, Supplement 2), S264–S272. https://doi.org/10.1016/j.amepre.2014.05.028

Barber, C. W., Ozonoff, V. V., Schuster, M., Hume, B. C., McLaughlin, H., Jannelli, L., & Saltzman, L. E. (1998). Massachusetts weapon-related injury surveillance system. *American Journal of Preventive Medicine, 15*(3), 57–66. https://doi.org/10.1016/S0749- 3797(98)00065-8

Barret, M. (2016). *HCUP external cause of injury (E-Code) evaluation report* (2016–03; HCUP Methods Series).

Baum, G. R., Baum, J. T., Hayward, D., & MacKay, B. J. (2022). Gunshot wounds: Ballistics, pathology, and treatment recommendations, with a focus on retained bullets. *Orthopedic Research and Reviews, 14*, 293–317. https://doi.org/10.2147/ORR.S378278

Borica, S. (2024). *Governor signs bill into law created by Lake County leaders that will speed up public release of crime data*. https://www.lakemchenryscanner.com/2024/08/14/governor-signs-bill-into-law-created-by-lake-county-leaders-that-will-speed-up-public-release-of-crime-data/

Braga, A. A. (2022). Gun violence is a public health crisis that needs more applied criminologists. *Criminology & Public Policy, 21*(4), 811–837. https://doi.org/10.1111/1745-9133.12608

Braga, A. A., & Cook, P. J. (2018). The association of firearm Caliber with likelihood of death from gunshot injury in criminal assaults. *JAMA Network Open, 1*(3), e180833. https://doi.org/10.1001/jamanetworkopen.2018.0833

Braga, A. A., Griffiths, E., Sheppard, K., & Douglas, S. (2021). Firearm instrumentality: Do guns make violent situations more lethal? *Annual Review of Criminology, 4*(1), 147–164. https://doi.org/10.1146/annurev-criminol-061020-021528

Brantingham, P. J., Quintana-Navarrete, M., Iliff, C., Uchida, C. D., & Tita, G. E. (2024). Situational and victim correlates of increased case fatality rates in Los Angeles shootings, 2005–2021. *Journal of Urban Health: Bulletin of the New York Academy of Medicine, 101*(2), 272–279. https://doi.org/10.1007/s11524-024-00845-z

CDC. (2024, July 5). *Funded data projects*. Firearm Injury and Death Prevention. https://www.cdc.gov/firearm-violence/php/funded-surveillance/index.html

Centers for Disease Control and Prevention, National Center for Injury Prevention and Control. Web-based Injury Statistics Query and Reporting System (WISQARS) [online]) [2022]. Available from www.wisqars.cdc.gov.

Conner, A., Azrael, D., Lyons, V. H., Barber, C., & Miller, M. (2019). Validating the national violent death reporting system as a source of data on fatal shootings of civilians by law enforcement officers. *American Journal of Public Health, 109*(4), 578–584. https://doi.org/10.2105/AJPH.2018.304904

Conner, A., Azrael, D., & Miller, M. (2019). Suicide case-fatality rates in the United States, 2007 to 2014: A nationwide population-based study. *Annals of Internal Medicine, 171*(12), 885–895. https://doi.org/10.7326/M19-1324

Cook, P. (1987). Robbery violence. *Journal of Criminal Law and Criminology, 78*(2), 357.

Cook, P. J. (1985). The case of the missing victims: Gunshot woundings in the National Crime Survey. *Journal of Quantitative Criminology, 1*(1), 91–102. https://doi.org/10.1007/BF01065250

Cook, P. J. (1991). The technology of personal violence. *Crime and Justice, 14*, 1–71. https://doi.org/10.1086/449183

Cook, P. J., Rivera-Aguirre, A. E., Cerdá, M., & Wintemute, G. (2017). Constant lethality of gunshot injuries from firearm assault: United States, 2003–2012. *American Journal of Public Health, 107*(8), 1324–1328. https://doi.org/10.2105/AJPH.2017.303837

Cook, P. J., & Roman, J. K. (2021). *Improving data infrastructure to reduce firearms violence*. NORC. https://www.norc.org/content/dam/norc- org/pdfs/Improving%20Data%20Infrastructure%20to%20Reduce%20Firearms%20Viole nce_Final%20Report.pdf

DC firearm injury surveillance dashboard| DC_FASTER. (2023). https://dchealth.dc.gov/dc_faster

Fackler, M. L., Bellamy, R. F., & Malinowski, J. A. (1988). The wound profile: Illustration of the missile-tissue interaction. *Journal of Trauma and Acute Care Surgery, 28*(1), S21.

FBI Criminal Justice Information Services (CJIS). (2022). *UCR program: Quarterly*. https://le.fbi.gov/file-repository/ucr-program-quaterly-october-2022.pdf/view

FBI releases 2022 crime in the nation statistics. (2022). [Press Release]. Federal Bureau of Investigation. https://www.fbi.gov/news/press-releases/fbi-releases-2022-crime-in-the-nation-statistics

Federal Bureau of Investigation. (2024). *Use-of-Force* [Folder]. Federal Bureau of Investigation. https://www.fbi.gov/how-we-can-help-you/more-fbi-services-and-information/ucr/use-of-force

Fowler, K. A., Leavitt, R. A., Betz, C. J., Yuan, K., & Dahlberg, L. L. (2021). Examining differences between mass, multiple, and single-victim homicides to inform prevention: Findings from the National Violent Death Reporting System. *Injury Epidemiology, 8*(1), 49. https://doi.org/10.1186/s40621-021-00345-7

Fox-Sowell, S. (2023, October 4). Minnesota's new crime data tool helps keep law agencies accountable, official says. *StateScoop*. https://statescoop.com/minnesota-crime-reporting-tool/

Gilbert, L., Parker, S., & Schechter, L. (2024). The impact of the COVID-19 pandemic on treatment for domestic violence injuries: Evidence from medical claims. *Review of Economics of the Household, 22*(2), 535–562. https://doi.org/10.1007/s11150-024- 09696-8

Gobaud, A. N., Mehranbod, C. A., Kaufman, E., Jay, J., Beard, J. H., Jacoby, S. F., Branas, C. C., Bushover, B., & Morrison, C. N. (2023). Assessing the gun violence archive as an epidemiologic data source for community firearm violence in 4 US cities. *JAMA Network Open, 6*(6), e2316545. https://doi.org/10.1001/jamanetworkopen.2023.16545

Goldstick, J. E., Cunningham, R. M., & Carter, P. M. (2022). Current causes of death in children and adolescents in the United States. *New England Journal of Medicine, 386*(20), 1955–1956. https://doi.org/10.1056/NEJMc2201761

Hargarten, S. W., Karlson, T. A., O'Brien, M., Hancock, J., & Quebbeman, E. (1996). Characteristics of firearms involved in fatalities. *JAMA, 275*(1), 42–45. https://doi.org/10.1001/jama.1996.03530250046025

HCUPnet data tools – Healthcare cost and utilization project (HCUPnet). (2022). https://datatools.ahrq.gov/hcupnet/

Hipple, N. K. (2022). Towards a National Definition and Database for Nonfatal Shooting Incidents. *J Urban Health.* Jun, 99(3), 361-372. doi: 10.1007/s11524-022-00638-2. Epub 2022 Apr 25. PMID: 35469106; PMCID: PMC9187788.

Hipple, N. K. (2022). Towards a National Definition and Database for Nonfatal Shooting Incidents. *J Urban Health.* Jun, 99(3), 361-372. doi: 10.1007/s11524-022-00638-2. Epub 2022 Apr 25. PMID: 35469106; PMCID: PMC9187788.

Houry, D., Sachs, C. J., Feldhaus, K. M., & Linden, J. (2002). Violence-inflicted injuries: Reporting laws in the fifty states. *Annals of Emergency Medicine, 39*(1), 56–60. https://doi.org/10.1067/mem.2002.117759

Hunt, P. R., Hackman, H., Berenholz, G., McKeown, L., Davis, L., & Ozonoff, V. (2007). Completeness and accuracy of International Classification of Disease (ICD) external cause of injury codes in emergency department electronic data. *Injury Prevention, 13*(6), 422–425. https://doi.org/10.1136/ip.2007.015859

ICD-10-CM official guidelines for coding and reporting. (2023). https://www.cms.gov/files/document/fy-2023-icd-10-cm-coding-guidelines-updated-01/11/2023.pdf

Kalesan, B., Adhikarla, C., Pressley, J. C., Fagan, J. A., Xuan, Z., Siegel, M. B., & Galea, S. (2017). The hidden epidemic of firearm injury: Increasing firearm injury rates during 2001–2013. *American Journal of Epidemiology, 185*(7), 546–553. https://doi.org/10.1093/aje/kww147

Karlson, T., & Hargarten, S. (1998). *Reducing firearm injury and death: A public health sourcebook.* https://scholar.google.com/scholar_lookup?title=Reducing+Firearm+Injury+and+Death%3A+A+Public+Health+Sourcebook+on+Guns&author=TA+Karlson&author=SW.+Hargart en&publication_year=1997&

Kaufman, E., Holena, D. N., Yang, W. P., Morrison, C. N., Jacoby, S. F., Seamon, M., Sims, C., Wiebe, D. J., & Beard, J. H. (2019). Firearm assault in Philadelphia, 2005–2014: A comparison of police and trauma registry data. *Trauma Surgery & Acute Care Open, 4*(1), e000316. https://doi.org/10.1136/tsaco-2019-000316

Kaufman, E. J., Song, J., Xiong, R., Seamon, M. J., & Delgado, M. K. (2024). Fatal and nonfatal firearm injury rates by race and ethnicity in the United States, 2019 to 2020. *Annals of Internal Medicine*. https://doi.org/10.7326/M23-2251

Kaufman, E. J., Wiebe, D. J., Xiong, R. A., Morrison, C. N., Seamon, M. J., & Delgado, M. K. (2021). Epidemiologic trends in fatal and nonfatal firearm injuries in the US, 2009–2017. *JAMA Internal Medicine, 181*(2), 237–244. https://doi.org/10.1001/jamainternmed.2020.6696

Kellermann, A. L., Rivara, F. P., Lee, R. K., Banton, J. G., Cummings, P., Hackman, B. B., & Somes, G. (1996). Injuries due to firearms in three cities. *New England Journal of Medicine, 335*(19), 1438–1444. https://doi.org/10.1056/NEJM199611073351906

Lauritsen, J. L., & Lentz, T. S. (2019). National and local trends in serious violence, firearm victimization, and homicide. *Homicide Studies, 23*(3), 243–261. https://doi.org/10.1177/1088767919848665

Lei, L., Goldstick, J. E., & Maust, D. T. (2023). Impact of firearm injury in children and adolescents on health care costs and use within a family. *Preventive Medicine, 175*, 107681. https://doi.org/10.1016/j.ypmed.2023.107681

Live crime tracker | NORC at the University of Chicago. (n.d.). Retrieved August 15, 2024, from https://www.norc.org/research/projects/live-crime-tracker.html

Ludwig, J., & Miller, J. (2024, April 29). *A lethal shift in America's gun violence crisis.* Governing. https://www.governing.com/policy/a-lethal-shift-in-americas-gun-violence-crisis

MacPhaul, E., Zhou, L., Mooney, S. J., Azrael, D., Bowen, A., Rowhani-Rahbar, A., Yenduri, R., Barber, C., Goralnick, E., & Miller, M. (2023). Classifying firearm injury intent in electronic hospital records using natural language processing. *JAMA Network Open, 6*(4), e235870. https://doi.org/10.1001/jamanetworkopen.2023.5870

Magee, L. A., Aalsma, M. C., Fortenberry, J. D., Gharbi, S., & Wiehe, S. E. (2022). Mental health outcomes from direct and indirect exposure to firearm violence: A cohort study of nonfatal shooting survivors and family members. *Journal of Criminal Justice, 83*, 101961. https://doi.org/10.1016/j.jcrimjus.2022.101961

Magee, L. A., Ortiz, D., Adams, Z. W., Marriott, B. R., Beverly, A. W., Beverly, B., Aalsma, M. C., Wiehe, S. E., & Ranney, M. L. (2023). Engagement with mental health services among survivors of firearm injury. *JAMA Network Open, 6*(10), e2340246. https://doi.org/10.1001/jamanetworkopen.2023.40246

Magee, L. A., Ortiz, D., Macy, J. T., Tolliver, S., Alvarez-Del-Pino, J., Kaur, A., Spivey, E., & Grommon, E. (2024). Identifying overlaps and disconnects between media reports and official records of nonfatal firearm injuries in Indianapolis, Indiana, 2021–2022. *Preventive Medicine, 180*, 107892. https://doi.org/10.1016/j.ypmed.2024.107892

Magee, L. A., Ranney, M. L., Fortenberry, J. D., Rosenman, M., Gharbi, S., & Wiehe, S. E. (2021). Identifying nonfatal firearm assault incidents through linking police data and clinical records: Cohort study in Indianapolis, Indiana, 2007–2016. *Preventive Medicine, 149,* 106605. https://doi.org/10.1016/j.ypmed.2021.106605

May, J. P., Hemenway, D., & Hall, A. (2002). Do criminals go to the hospital when they are shot? *Injury Prevention, 8*(3), 236–238. https://doi.org/10.1136/ip.8.3.236

Miller, K. R., Egger, M. E., Pike, A., Burden, J., Bozeman, M. C., Franklin, G. A., Nash, N. A., Smith, J. W., Harbrecht, B. G., & Benns, M. V. (2022). The limitations of hospital and law enforcement databases in characterizing the epidemiology of firearm injury. *Journal of Trauma and Acute Care Surgery, 92*(1), 82–87. https://doi.org/10.1097/TA.0000000000003367

Miller, M., Azrael, D., Yenduri, R., Barber, C., Bowen, A., MacPhaul, E., Mooney, S. J., Zhou, L., Goralnick, E., & Rowhani-Rahbar, A. (2022). Assessment of the accuracy of firearm injury intent coding at 3 US hospitals. *JAMA Network Open, 5*(12), e2246429. https://doi.org/10.1001/jamanetworkopen.2022.46429

National electronic injury surveillance system—Firearm injury surveillance study (NEISS-FISS)—Healthy People 2030 | health.gov. (2024). https://health.gov/healthypeople/objectives-and-data/data-sources-and-methods/data-sources/national-electronic-injury-surveillance-system-firearm-injury-surveillance-study-neiss-fiss

NEISS coding manual. (2019). https://www.cpsc.gov/s3fs-public/January-2021-NT-CPSC-only-NEISS-Coding-Manual.pdf?xa_nMM1kB4SGpuSMOwf0NHkkkIqNcn8F

NIBRS technical specification. (2023). [File]. Law Enforcement. https://le.fbi.gov/file-repository/nibrs-technical-specification-063023.pdf/view

Nix, J., & Shjarback, J. A. (2021). Factors associated with police shooting mortality: A focus on race and a plea for more comprehensive data. *PloS One, 16*(11), e0259024. https://doi.org/10.1371/journal.pone.0259024

NORC. (2020). *The state of firearms data in 2019.* https://www.norc.org/PDFs/Firearm%20Data%20Infrastructure%20Expert%20Panel/Stat e%20of%20Firearms%20Research%202019.pdf

Office of the U.S. Surgeon General. (2024). *Firearm violence in America: A public health crisis* [Text]. U.S. Office of the Surgeon General. https://www.hhs.gov/surgeon-general/priorities/firearm-violence/index.html

Parker, S. T. (2022). Measuring gun violence in police data sources: Transitioning to NIBRS. *Injury Epidemiology, 9*(1), 15. https://doi.org/10.1186/s40621-022-00376-8d

Real-Time Crime Index. (n.d.). *AH Datalytics.* Retrieved August 15, 2024, from https://www.ahdatalytics.com/rtci/

Roman, J. K., Washburn, A., Rodriguez, S., Roman, C. G., Navarro, E. I., Brey, J. T., & Reist, B. M. (2023). *Harms after a victimization: Experience and needs (HAVEN)* (307781). https://www.norc.org/research/projects/harms-after-a-victimization-experience-and- needs-haven.html

Schumacher, S., Kirzinger, A., Presiado, M., Valdes, I., & Published, M. B. (2023, April 11). Americans' experiences with gun-related violence, injuries, and deaths. *KFF.* https://www.kff.org/other/poll-finding/americans-experiences-with-gun-related-violence- injuries-and-deaths/

Song, Z., Zubizarreta, J. R., Giuriato, M., Koh, K. A., & Sacks, C. A. (2023). Firearm injuries in children and adolescents: health and economic consequences among survivors and family members. *Health Affairs*, *42*(11), 1541–1550. https://doi.org/10.1377/hlthaff.2023.00587

Stewart, C. C., Simon, G., Ahmedani, B. K., Beck, A., Daida, Y. G., Lynch, F. L., Owen-Smith, A. A., Negriff, S. L., Rossom, R., Sterling, S. A., Lu, C. Y., & Schoenbaum, M. (2024). Variation in completeness of coding external cause of injuries under ICD-10-CM. *Injury Prevention*, ip–2023–045164. https://doi.org/10.1136/ip-2023-045164

Swanson, J. W., Zeoli, A. M., Frattaroli, S., Betz, M., Easter, M., Kapoor, R., Knoepke, C., Norko, M., Pear, V. A., Rowhani-Rahbar, A., Schleimer, J. P., & Wintemute, G. J. (2024). Suicide prevention effects of extreme risk protection order laws in four states. *The Journal of the American Academy of Psychiatry and the Law*, JAAPL.240056-24. https://doi.org/10.29158/JAAPL.240056-24

Tessler, R. A., Arbabi, S., Bulger, E. M., Mills, B., & Rivara, F. P. (2019). Trends in firearm injury and motor vehicle crash case fatality by age group, 2003–2013. *JAMA Surgery*, *154*(4), 305–310. https://doi.org/10.1001/jamasurg.2018.4685

Traci Ramirez on behalf of NCHS's ICD-10-CM Classification Team. (2024, August 5). [Personal communication].

UCR Program Quarterly: October 2022 (22-2). October 2022 issue of the UCR Program Quarterly, issued by the Criminal Justice Information Services (CJIS). Available: https://le.fbi.gov/file-repository/ucr-program-quaterly-october-2022.pdf/view

Vyrostek, S. B., Annest, J. L., & Ryan, G. W. (2001). *Surveillance for fatal and Nonfatal injuries—United States, 2001*. https://www.cdc.gov/mmwr/preview/mmwrhtml/ss5307a1.htm

Ward, J. A., Cepeda, J., Jackson, D. B., Johnson, O., Webster, D. W., & Crifasi, C. K. (2024). National burden of injury and deaths From shootings by police in the United States, 2015–2020. *American Journal of Public Health*, *114*(4), 387–397. https://doi.org/10.2105/AJPH.2023.307560

WISQARS nonfatal injury reports. (n.d.). Retrieved March 26, 2019, from https://webappa.cdc.gov/sasweb/ncipc/nfirates.html

Zimring, F. E. (1972). The medium Is the message: Firearm caliber as a determinant of death from assault. *The Journal of Legal Studies*, *1*(1), 97–123. https://doi.org/10.1086/467479

2

Police Investigation and Court Processing of Shootings in Durham, NC

Philip J. Cook and Audrey Vila

2.1 Overview

Gun violence is the most costly crime problem in many American cities (Cook and Ludwig 2022). Police departments have a leading role in gun-violence prevention, primarily through their capacity to investigate criminal shootings, arrest perpetrators, and provide the courts with evidence required for a successful prosecution. Their performance in accomplishing these tasks has a direct influence on gun violence rates and is important as well in serving victims and their communities by delivering appropriate punishment to perpetrators. In short, the effectiveness of police investigations and subsequent prosecutions is directly relevant to both prevention and justice.

Monitoring the performance of the police in responding to gun violence requires measures that are directly related to prevention and justice. The primary mechanisms by which police investigations, and more generally the criminal justice system, prevent gun violence include deterrence of would-be shooters by creating a credible threat of punishment, and incapacitation of violence-prone individuals who have been convicted (Chalfin and McCrary 2018; Nagin 2013; Cook 1980; Zimring and Hawkins 1973). Both of these mechanisms require that culpable shooters are identified, arrested, convicted, and punished (Abrams 2012; Johnson and Raphael 2012). Data appropriate to measuring performance then requires linking incidents to police-investigation outcomes and, ultimately, the disposition of resulting cases in court. Similar data are also relevant to assessing the delivery of justice.

In practice linked data of this sort are not routinely available and rarely compiled. The available performance measures for the police are versions of arrest or clearance rates, with no indication of the ultimate disposition (Cook and Mancik 2024). The result is that it is usually not possible to measure the

DOI: 10.1201/9781003430773-3

performance of the criminal justice system using a suitable endpoint, such as the likelihood that a shooter is convicted and sentenced to prison. We offer the current analysis to demonstrate what is required to produce such measures, and what can be learned when they are available. Our analysis was made possible through a close working relationship with the Durham Police Department (DPD), which has an exemplary record-keeping system.

One finding from these data is that many arrests do not result in a conviction, and hence do not contribute to the deterrence and incapacitation mechanisms, or the delivery of justice. During the five-year period under consideration, there were 1188 shooting victims in the city of Durham (of which 160 died) in 1011 distinct incidents – about 200 incidents per year on average, in a city of about 284,000 residents. As a result of DPD investigations, 46% of the fatal shooting incidents resulted in at least one arrest, compared with just a 16% arrest rate for incidents in which the victim(s) survived. The arrestees were processed in the Durham courts (although a handful of cases were transferred to federal court). Disposition data were available through July 2022, or, in the case of FS cases, through November 2022. While most defendants were prosecuted, some had all charges dismissed for various reasons. The remaining cases – those in which at least one defendant was prosecuted – represent 42% of the fatal shooting incidents and just 11% of the nonfatal shooting incidents. For all 1011 shooting incidents combined, there were 74 with at least one conviction in the Durham courts, 6 that were federally indicted, and 10 that were processed as juveniles (so that we lacked access to court records). Another 62 cases were still pending as of July 2022. If we assume that all of the federal cases and the cases pending in Durham courts eventually result in a conviction, the fraction of shooting incidents in which there was both an arrest and conviction was about 15%. Thus when there is a criminal shooting in Durham, it is unlikely (85%) that anyone is ultimately arrested and convicted.

These results are troubling, not only because the system has failed to serve the survivors and their community by delivering justice (Forst 2024; Friedorsdof 2021) but also because of the urgent need to prevent future gun violence (Anderson 1999; Brunson & Wade 2019; Daly 2023). As noted, an increase in the arrest and conviction rates for gun-violence cases would reduce future gun violence by incapacitating dangerous perpetrators from engaging in subsequent violence and deterring some would-be shooters by communicating a credible threat that shooting someone will result in legal consequences. Other mechanisms may reinforce this prevention effect. For example, in some cases, placing the shooter in custody will interrupt cycles of revenge. Further, it is plausible that a stronger performance by the police and courts would serve to improve community cooperation with police investigations by sending a clear message that the authorities are determined to help improve the quality of life in violence-impacted communities (Braga and Cook 2023; Altholz 2020; Keel et al. 2009; Leovy 2015; Ousey & Lee 2008).

An additional finding is that there are large disparities between fatal and nonfatal shooting incidents with respect to how successfully they are investigated by the police and processed in court. Nonfatal shootings are sometimes called "almoscides" to highlight the fact that whether the victim lives or dies in a criminal shooting is largely a matter of chance (Leovy 2015; Braga and Cook 2018; Zimring 1968, 1972). The mixes of circumstances, motives, and characteristics of victims and shooters are similar. For that reason, it is reasonable to assert that solving nonfatal shooting cases is as important for prevention purposes as solving fatal shooting cases. The goal is to prevent shootings, period. Yet despite this logic, the police in Durham, like in other jurisdictions, are much more likely to make an arrest in fatal than nonfatal shootings, and that is true despite the fact that nonfatal shootings generally have a key witness (the victim) who is lacking from fatal cases (Cook et al. 2019). Much of the explanation for this disparity appears to be with respect to the greater priority and resources devoted to the investigation of fatal shootings. We are able to demonstrate that the gap widens still further when arrest data are coupled with data on court dispositions. In particular, the prosecutorial dismissal rates for arrests in nonfatal-shooting cases are higher than for fatal shootings.

Taken together these findings reinforce the idea that arrest is not enough for the purposes of prevention and justice, and arrest rates provide an incomplete performance measure for police and the courts. Statistics on the likelihood that an incident will result in at least one conviction provide a more complete accounting of performance with a relevant endpoint. But the slow pace of court processing implies that such statistics will not be knowable for any one period's crimes until several years later. Nevertheless, results of this sort should be readily available to local government agencies and the public. But compiling these data has not been easy and has only been possible due to extraordinary efforts on the part of the Durham Police and District Attorney's Office. We encourage police and court officials to coordinate on producing a similar report on an annual basis. The next section provides a brief description of data sources and processing.

2.1.1 Data Sources and Processing

The analysis here is based on data from the Durham Police Department (DPD) and the Durham Courts for incidents occurring during the five calendar years, 2017–2021. An incident was included if at least one victim was shot in the course of a criminal assault. In any one incident there may be more than one victim, as is often true, for example, in a drive-by shooting. Most of the analysis which follows tabulates incidents rather than individual victims, using the following definitions:

> Fatal Shooting Incident (FS): One or more gunshot victims, of whom at least one dies

Nonfatal Shooting Incident (NFS): One or more gunshot victims, all survive

The relevant DPD records were first extracted in February 2022. These data included descriptive information on the incident, the victim(s), and the results of the DPD investigation. These data were cleaned in consultation with the DPD's Crime Analysis Unit, and the result is a complete list of incidents known to the police that meet the criteria for inclusion. Note that most gun crimes, including criminal assaults and robberies, are excluded from these data because the victim was not shot.

The status of DPD's investigation was updated through July 15, 2022. Of the 1011 incidents studied, a total of 206 resulted in one or more arrests. These arrests were linked to court cases partly by use of docket numbers; these numbers are assigned to charges brought by the prosecution, and link to public records on court processing. Since early 2018, the majority of docket numbers are included in DPD records. For earlier arrests the docket numbers were found by use of other information available in the DPD records. Data on case processing in the courts was accessed from public records and merged with DPD data. In addition, in November 2022 the Office of the Durham District Attorney provided complete case-processing data for all FS cases originating from our sample of shooting incidents.

The resulting data set is unique in incorporating the entire trajectory from shooting incident to court disposition.

2.2 Patterns of Victimization

Here we tabulate the number of gunshot victims by year and provide descriptive information about these victims and incidents for the five years combined. Table 2.1 provides annual counts. Of the 1188 shooting victims, 160 were killed, for an overall fatality rate of 13%. The number of gunshot victims is highly variable from year to year, with an unprecedented

TABLE 2.1

Annual Count of Victims, by Outcome

Year	FS Victims	NFS Victims	All Victims
2017	23	194	217
2018	26	174	200
2019	33	156	189
2020	33	283	316
2021	45	221	266
Total	160	1028	1188

increase from 189 to 316 (67%) between 2019 and 2020. It should be noted that Durham's experience was replicated in many other jurisdictions, and the nationwide increase in gun violence from 2019 to 2020 was the highest in recorded history, with the firearms homicide rate surging by 35% (Kegler et al. 2022). It was a year that was marked by the profound disruptions and unrest associated with the onset of the COVID-19 pandemic and the nationwide demonstrations against the police following the death of George Floyd (Braga and Cook 2023).

TABLE 2.2

Annual Count of Shooting Incidents, Classified by Whether Any Victims Died

Year	FS Incidents	NFS Incidents	All Incidents
2017	23	170	193
2018	26	143	169
2019	31	128	159
2020	33	239	272
2021	41	177	218
Total	154	857	1011

Table 2.2 covers the same period but provides counts of incidents rather than individual victims. Two victims were shot and killed in six of the FS incidents, which accounts for the difference in totals (154 incidents, 160 victims). In addition, there were 38 surviving victims in FS incidents. Overall, there were 1011 shooting incidents, of which 13% had more than one gunshot victim. In 26 cases, 3% of the total, there were three or more gunshot victims.

A graph of incident counts by quarter (Figure 2.1) depicts the extraordinary surge in shooting incidents during 2020. Much of the surge was confined to the third quarter, July–September. During that period Durham was averaging more than one shooting incident per day. Total shooting incidents dropped between 2020 and 2021, although the number of deaths increased.

With respect to "who, when, and where," the shooting patterns tend to be fairly consistent over time. Table 2.3 displays data on victim characteristics for the five years combined. Gunshot victims in Durham were overwhelmingly male, African American, and youthful (age 18–34). The majority of Durham's residents are White, but only one in seven victims were White. (Most of these White victims were Hispanic.) With respect to season, rates were somewhat lower in winter than the rest of the year.

The comparison of fatal and nonfatal shootings is of interest in considering the priority of these two types of cases in allocating police investigation resources. To a considerable extent, the difference between life and death in a criminal shooting is a matter of chance (Zimring 1972; Braga and Cook 2018). Hence, it is not surprising that there are similar patterns of FS and NFS cases. There are exceptions – Table 2.3 indicates that a larger share of fatal shootings (two-thirds) are outdoors than nonfatal shootings (just about half),

FIGURE 2.1
Quarterly Counts of Shooting Incidents (2017–2021).

TABLE 2.3
Shooting Victims, 2017–2021 Combined: Who, When, Where

		FS Victims *N* = 160 %	NFS Victims *N* = 1028 %
Sex	Female	10.6	16.8
	Male	89.4	83.2
Age group	0–17	9.4	12.8
	18–24	30.0	31.9
	25–34	33.8	30.6
	35+	26.3	24.2
Race/ethnicity	White	14.4	13.0
	Black	83.8	86.1
	Other	1.9	0.9
	Hispanic	8.8	9.0
Location	Residence	25.0	35.9
	Indoor (non-residence)	7.5	12.2
	Outdoor	66.9	48.7
	Other/unknown	0.3	3.2
Quarter of the year	Q1	20.0	18.6
	Q2	26.3	22.3
	Q3	26.9	32.0
	Q4	26.9	27.1
Days	Weekend (Saturday and Sunday)	34.4	31.3
	Weekend (Friday, Saturday, Sunday)	45.0	45.0
	Weekend (Saturday, Sunday, Monday)	50.6	47.8

suggesting that in Durham, outdoor shootings tend to be more lethal than indoor. In other respects, the case mixes for fatal and nonfatal shootings are very similar. There is every reason to believe that the investigation of nonfatal shootings is as important from a prevention viewpoint as investigation of fatal shootings (Cook 2020).

2.3 Investigations

Over the five-year period 2017–2021, the Durham Police Department investigated an average of 202 incidents a year in which at least one victim was shot. That average conceals a wide variation, from 159 (in 2019) to 272 (in 2020). In general, the resources devoted to the investigation are greatly influenced by whether the victim dies. In particular, homicide investigators have lighter caseloads than the investigators who are assigned nonfatal cases (Cook, Ho and Shilling 2017). One result, in Durham as in other cities, is that investigations of FS cases are more likely to be successful (Cook, Braga et al. 2019).

The usual measure of success is a clearance rate. A case is considered "cleared" if there is at least one arrest, or there are exceptional circumstances in which the investigator has identified a suspect but is unable to make an arrest for some reason (Cook and Mancik 2024). For the five years combined, the arrest rate for FS incidents was 46%, while the clearance rate (including exceptional clearances) was 50%. For NFS incidents the arrest rate was just 16%, with a 21% overall clearance rate. Thus, the likelihood of investigative "success," measured either by arrest or by clearance, was much higher in FS than NFS incidents.

Most of the FS clearances are by arrest (71/77), but exceptional clearances are more common for NFS cases. Table 2.4 provides the explanations given for different types of clearances as of July 2022.

The focus in what follows is on clearance by arrest, since it is arrest that is the precondition for a successful outcome to the investigation. Of particular concern are the exceptional clearances in which the prosecution declined to accept for prosecution (presumably because the evidence was judged inadequate to support an indictment) or the victim refused to cooperate. In those cases, the shooter remained at large, so the investigation did not contribute to the prevention of future gun violence. However, one category of "exceptional clearance" does result in arrest and prosecution. We classify the handful of exceptional cases in which the suspect was taken in custody in another jurisdiction as "cleared by arrest." These are typically the result of a Durham warrant being served by another agency and hence should be considered a successful outcome for the DPD investigation.

TABLE 2.4

Type of Clearance, 2017–2021 Combined, for FS and NFS Incidents (Counts)

Case Status	FS Cases	NFS Cases
Cleared by Arrest	71	135
Exceptional clearance – "Prosecution declined"	2	15
Exceptional clearance – "Victim refused to cooperate"	0	32
Exceptional clearance – NA	4	0
Inactive/Open	77	675
Total	154	857

2.3.1 Focus on Arrests

Figure 2.2 illustrates the annual rates of arrest for the two groups of incidents. The large gap between arrest rates for FS and NFS cases appears each year. Also evident is that FS cases during the first three years had higher arrest rates than in 2020–2021, the COVID years.

A case is counted as cleared by arrest if there is at least one arrest, regardless of the number of perpetrators in the incident. It is not uncommon for a successful investigation to result in more than one arrest, as shown in Table 2.5. That was true in 28% of the successful FS investigations and 18% of the successful NFS investigations. One FS incident resulted in six arrests in all.

It is also of interest to observe the timing of the (first) arrest in these investigations. Table 2.6 and Figure 2.3 depict the time to first arrest for FS and NFS Incidents. The time to arrest is an indicator of what is known as the "solvability" of the case (Cook and Lopez 2024). Some cases are readily solved, and sometimes result in an arrest on scene or shortly thereafter. An extreme example would be a domestic shooting in which the perpetrator confesses to the first officer on the scene. At the other end of the spectrum are drive-by shootings involving gang conflicts in which victims and other witnesses are unlikely to cooperate. Solving such cases, if they are solved, may require the patient collection and processing of digital and physical evidence, and "converting" reluctant witnesses. In short, the more difficult cases require resources and time, which may be lacking, especially in lower-priority non-fatal shootings.

Several patterns are evident:

- For FS cases, the arrest rate dropped below 50% in 2019, with a much steeper drop in 2020 down to 30%.
- At every point following the shooting, the likelihood that there has been at least one arrest is substantially higher for FS incidents than NFS incidents.
- A notable portion of all arrests occurs within 2 days. That includes 25% (18/71) of successful FS investigations and 34% (46/135) of NFS

successful investigations. The median time until arrest for successful investigations is 12 days for FS incidents and 9.5 days for NFS incidents.

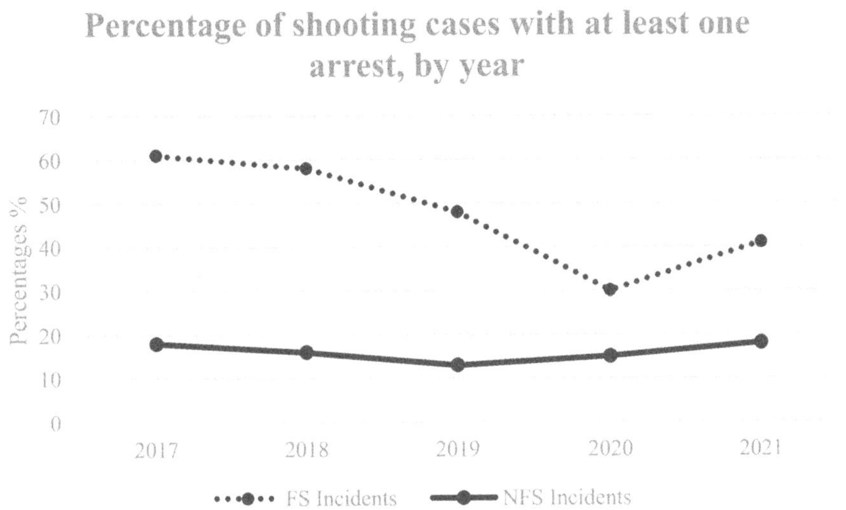

FIGURE 2.2
Percentage of Shooting Cases with at least One Arrest, by Year.

TABLE 2.5

Number of Arrests per Successful Investigation, 2017–2021 Combined (Counts)

Number of Arrests	FS Incidents	NFS Incidents
1	51	111
2	11	15
3	6	8
4 or more[a]	3	1
Total cases with arrests	71	135

[a]*One FS case resulted in six arrests. Others in this category had 4.*

What explains the sharp drop of arrest rates in 2020? It should be noted that Durham's FS arrest rate in 2018–2019 was in line or even higher than national rates, and that the national rate also dropped sharply in 2020 (Cook and Mancik 2024). That year DPD (and the nation) was faced with a surge in its homicide caseload, at the same time that the number of sworn officers was declining in the face of nationwide protests against the police. Evidence strongly suggests that the caseload of homicide investigators has a direct (negative) effect on arrest rates (Mancik and Parker 2019; PERF 2019; Cook

TABLE 2.6

Current Status in Durham Courts of FS Cases from 2017 to 2021 Incidents Combined (Counts)

	Pending	Convictions	Dismissals with No Federal Indictment	Federally Indicted	Totals
Incidents	35	27	8	2	72[a]
Individuals	50	33	9	9	101
Charges	81	49	14	81	225

[a]*While only 71 FS incidents were cleared by arrest, there was a felony warrant outstanding for an additional incident.*

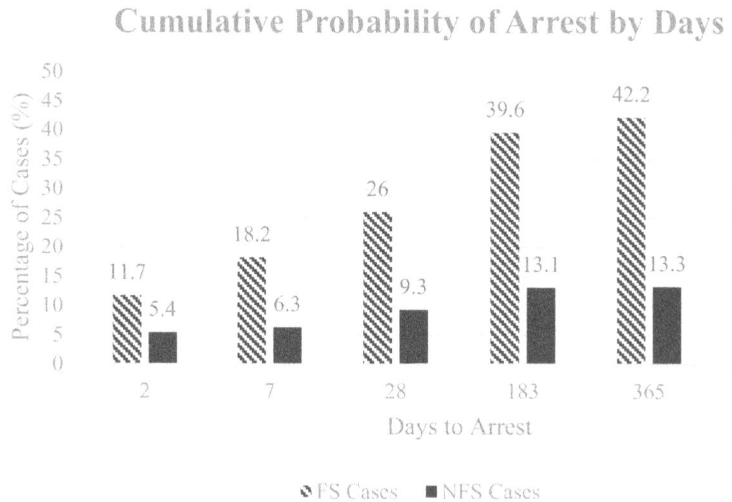

FIGURE 2.3
Cumulative Probability of Arrest by Days. Note: The "Ever" statistics include arrests in other jurisdictions for which there was no date of arrest in the DPD data.

et al. 2019). It is also plausible that civilian cooperation with police investigations declined in 2020.

Why are FS investigations more likely to be successful than NFS investigations? One plausible explanation is in terms of the resources devoted to the investigation (Cook et al. 2019). The homicide investigators generally have lighter caseloads and greater access to other resources. Investigations tend to be more sustained, as suggested by the difference in median time to arrest. Another plausible explanation is that NFS investigations are often stymied by the challenge of gaining the cooperation of the victim, without which it may be essentially impossible to make a case against a suspect that would be sufficient to gain a conviction in court (Braga and Cook 2023; White et al.

2021; Brunson et al. 2019; Cook, Ho, and Shilling 2017). (Of course FS investigations are rarely able to interview the victim, but at least the victim does not interfere with a successful investigation, and other members of the community tend to be more forthcoming if the victim has died.)

2.3.2 Evidence Collection

Generally speaking, more evidence is collected in FS investigations than in NFS investigations (Cook et al. 2019). Figure 2.4 reports the prevalence of some of the common types of evidence for the five-year period combined. These results are suggestive of the greater resource intensity of FS investigations. Another indicator is the number of attachments included in the DPD investigation files, covering such areas as investigator notes, warrants and waivers, audio and video files, and so forth. The median FS investigation included 16 such attachments, compared with just five for NFS investigations.

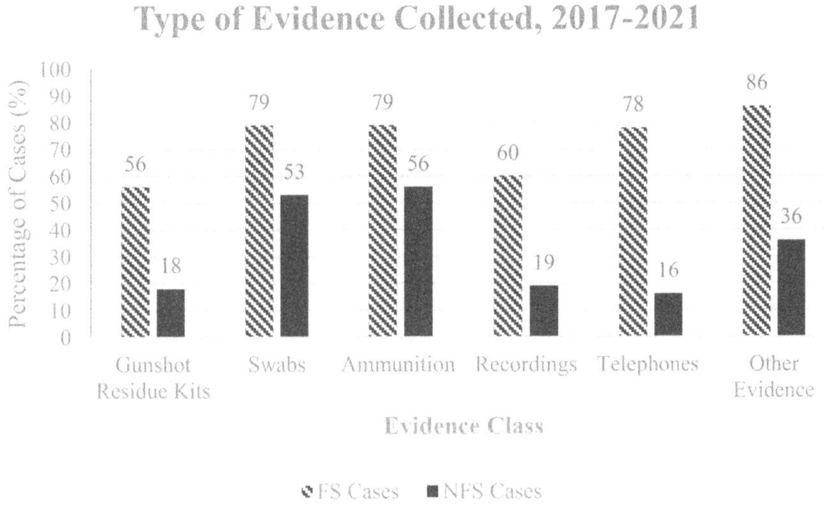

FIGURE 2.4
Type of Evidence Collected , 2017–2021.

2.4 Court-Processing Outcomes

While the usual performance measure for police investigations is the arrest rate or clearance rate, it is important to observe what happens after arrest. In theory, the prevention goal of police investigation is accomplished primarily through incapacitation and deterrence (Chalfin and McCrary 2017).

Deterrence requires a credible threat of legal consequences for crime, while incapacitation requires restraint of dangerous people. Arrest only contributes to these mechanisms to the extent that it results in conviction and punishment. Yet in practice a nontrivial percentage of arrests result not in conviction, but rather a dismissal of all charges. There is also a possibility that a defendant will be acquitted following trial, but that is a rare event.[1]

In what follows we analyze the case histories for FS cases first, and then NFS cases. Recall that an FS case (or incident) is one in which at least one gunshot victim died. NFS cases are those in which one or more victims were shot, but none died.

To be clear, in our usage, the terms "incident" and "case" are closely related. The "incident" is the original shooting event, which results in one or more people killed or wounded. The "case" includes all arrests, defendants, and charges stemming from a particular incident.

2.4.1 FS Cases

A total of 71 FS incidents occurring during the period 2017–21 resulted in at least one arrest; there was a felony warrant outstanding for one other incident. Note that in a number of these incidents more than one individual was arrested and prosecuted, typically on more than one charge. Overall, the 72 cases included more than 101 defendants and 225 charges.[2] Table 2.6 provides the status of these cases, defendants, and charges as of July 2022. Cases are classified as follows:

- "Pending" if at least one defendant is still under indictment, and none have been convicted, as of November 2022.
- "Conviction" if at least one defendant has been convicted by then.
- "Dismissal" if charges against all defendants have been dismissed.

Thus, as of November 2022, 27 of the 72 cases (38%) had resulted in any conviction in the Durham court, and half (49%) were pending. In the remaining 10 cases (14%), all charges against all defendants were dismissed – although in two of those cases, the reason for dismissal was a transfer to federal court. (We lack data on the disposition of those cases.) Most of the other cases were dismissed for lack of evidence. The means of conviction (not shown in the table) was in every case but one a guilty plea.

Table 2.7 divides case dispositions by the year of the incident. Most cases originating from the gun murders ("FS incidents") from 2020 and 2021 were still pending as of November 2022, with the exception of one that was dismissed and three that resulted in conviction. (It should be noted that court operations were impaired by the restrictions implemented as a result of the COVID-19 pandemic.) Focusing on the oldest cases, from incidents occurring

TABLE 2.7

November 2022 Status of FS Cases by Year of the Shooting (Counts)

Incident Year	Pending	Convictions	Dismissed	Federally Indicted	Totals
2017	2	8	4	0	14
2018	1	12	1	1	15
2019	8	4	2	1	15
2020	8	2	1	0	11
2021	16	1	0	0	17
Totals	35	27	8	2	72[a]

[a]*While only 71 FS incidents were cleared by arrest, there was a felony warrant outstanding for an additional incident.*

in 2017 and 2018, we see that 20 of the 29 resulted in at least one conviction and another was federally indicted.

TABLE 2.8

Most Serious Conviction Charge for FS Cases that Resulted in at least One Conviction

Felony Class[a]	Count
A	1
B1	16
B2	1
C	2
D	4
F	2
H	1

[a] *"Felony class" is the class of the most serious charge of conviction. In cases in which more than one defendant was convicted, "felony class" is the class of the convicted charge for the defendant who was convicted of the most serious charge.*

The plea-bargaining process is evident from the gap between the initial charge and the ultimate conviction charge. Most telling is that while 21 of the convicted defendants were initially charged with an A felony (which carries a sentence of life without parole), the only one that was convicted at that level was the lone FS defendant convicted at trial.[3] Of the 27 FS cases in which there has been at least one defendant convicted, the most serious charge for 16 cases was a B1 felony, which has a minimum presumptive sentence of 16–20 years in prison, and more if the defendant has a prior record of convictions. Seven others were convicted of a B2, C, or D felony, in which an active prison sentence is presumptive, and the other three of an F or H felony in which only defendants who have a serious record of prior convictions presumptively receive active sentences. Note that several of these defendants were also convicted of other charges, and the judge had the possibility of making the sentences consecutive. In any event, we lack information about the actual sentences (Table 2.8).

2.4.2 *NFS* Cases

Of the 135 incidents in which there was at least one arrest recorded in the DPD data, we were able to track 129 cases in the Durham courts. As shown in Table 2.9, 47 resulted in at least one conviction, all by guilty plea, while 27 are pending. In cases stemming from 45 incidents, all charges were dismissed against all defendants; in four of those incidents (involving six defendants and 23 charges), the defendants were federally indicted. Among the 41 dismissals without federal indictment, there were six incidents in which the defendant was prosecuted for a different, even more serious criminal incident. In the other 35 incidents, all defendants went free.

TABLE 2.9

Current Status of NFS Cases (Counts)

	Juveniles	Pending	Convictions	Dismissals with No Federal Indictment[a]	Federally Indicted	Totals
Incident #s	10	27	47	41	4	129[b]
Individuals	12	32	49	43	6	142
Charges	29[c]	93	83	88	23	316

Note: The previous footnote also applies to this table.

[a]*These dismissal counts do not include cases where the local court dismissed charges in response to federal indictments. In six of these incidents the charges were dismissed for this shooting, but the defendants were prosecuted for another crime that was deemed more serious.*

[b]*There were six incidents that had arrests, but we were unable to locate docket information (for more details, see explanatory memo).*

[c]*Juvenile charges are based on police data and cannot be confirmed from court records. Therefore, there may be more or fewer charges depending on what the DA's office decided to charge.*

Table 2.10 groups cases according to the year of the incident. Most of the pending cases (22 of 27) are from 2020 and 2021. The conviction percentage for 2017–2019 incidents, where there are only five pending cases, is 47%. If all the pending cases eventually result in a conviction, and the cases that were dismissed due to federal indictment result in conviction in federal court,

TABLE 2.10

Current Status of NFS Cases by Year of Incident (Counts)

Incident Year	Juveniles	Pending	Convictions	Dismissed	Federally Indicted	Totals
2017	1	2	16	9	1	29
2018	0	2	9	10	2	23
2019	2	1	7	5	1	16
2020	5	7	10	11	0	33
2021	2	15	5	6	0	28
Totals	10	27	47	41	4	129

then the percentage of incidents with at least one conviction would be about 63% (excluding juveniles).

For the entire five-year period, there were 49 defendants convicted in the 47 cases in which there was at least one conviction. We do not have information on sentencing for these defendants.

2.4.3 *Comparison of FS and NFS*

The long delays in processing FS cases make it difficult to estimate the number of cases that resulted in at least one conviction. But since dismissals tend to occur soon after indictment (if at all), it is reasonable to assume that the tables above include all but a handful of cases that were ultimately dismissed by the district attorney. Table 2.11 provides the "bottom line" with respect to dismissal rates, which are three times as high for NFS as FS cases. One result is that the gap in arrest rates between FS and NFS cases is widened when dismissals are considered.

Judging from the results for 2017 and 2018 cases, it is reasonable to presume that most of the remaining cases will eventually result in a conviction. It is rare indeed for a defendant to be judged not guilty following prosecution.

TABLE 2.11

Dismissal Rates Overall

	FS	NFS
Dismissal rate	8/72 = 11.1%	41/119 = 34.5%

Note: Dismissals by Durham courts in conjunction with a federal indictment are not counted in computing this rate. Juvenile cases are excluded from both the numerator and denominator of the NFS calculation.

2.5 Conclusions

Gun violence is arguably Durham's most pressing crime problem, as it is for many other American cities (Cook and Ludwig 2022). There is every reason to believe that the problem is exacerbated by the fact that so few shootings result in arrest and conviction. This chapter has documented that fact in some detail, and in particular provides statistics on conviction rates that have not been publicly available before.

In brief, for incidents occurring during the five-year period 2017–2021, 27 of the 154 fatal-shooting incidents 18% resulted in a conviction by November 2022, and defendants for two other cases were indicted in federal court. Only 47 of the 857 nonfatal shooting incidents (5%) resulted in a conviction.[4] If we confine the calculation to incidents for the years 2017–2019 (to allow most cases the time to have worked their way through the courts), the percentages increase to 33% for FS incidents and 8% for nonfatal incidents (assuming that the federal cases resulted in conviction). Taking all shooting cases together for 2017–19, 17% have resulted in at least one arrest and 12% in conviction.[5] The convictions do not necessarily result in an active prison sentence, even for the FS cases. All but one conviction resulted from a plea bargaining process rather than a trial.

Regardless of which statistics are used, two substantive conclusions stand out. First, people who shoot others in criminal assault incidents are unlikely to be arrested and convicted. The chances of "getting away with it" are seven in eight. Second, there is a large disparity in conviction rates depending on whether the victim lives or dies; the likelihood of arrest followed by conviction is four times as high if the victim dies. We have learned by including court processing data that the disparity in conviction rates is still larger than the disparity in arrest rates.

What can be done to improve the effectiveness of law enforcement against gun violence, and in particular improving the performance of investigations and prosecutions (Wellford et al. 2019; Braga and Dusseault 2018)? A good place to start is increased resources and an evidence-based plan for using them most effectively (Bureau of Justice Assistance 2018; PERF 2019). Creating a plan of this sort is well beyond the scope of the current chapter, but here are two relevant observations.

First, in Durham and elsewhere one pervasive obstacle to successful investigation and prosecution is the lack of witness cooperation. While the police have access to increasingly sophisticated technology for obtaining forensic evidence, human witnesses remain essential to solving crimes and trying cases. There are a variety of ways in which the police and courts could reduce the barriers to cooperation: protecting against retaliation, ensuring that financial costs of cooperation including lost wages are reimbursed, and – perhaps most important – taking the time and effort required to overcome witnesses' distrust of the police. These and other approaches require additional resources.

Second, while it is understandable that police, prosecutors, and courts devote more resources to murders than to nonfatal shootings, it is hard to justify that disparity if the goal is preventing future gun violence. There is a large element of chance in whether a shooting victim lives or dies. As a logical consequence, the mix of fatal cases is similar to the mix of nonfatal cases. In particular, substantial overlap exists in the mix of circumstances, motivations, and people involved. Reducing the rate of future shootings would inevitably reduce the rate of gun murders, and that can be accomplished by improving success rates in investigating and prosecuting nonfatal shootings as well as fatal.

Notably, the Denver Police Department has created a model program with promising results that includes a well-resourced investigative unit dedicated to nonfatal shooting cases. The goal is that these cases should have the same investigative priority as murders (Phillips et al. 2022). The problems in Durham reflect the intrinsic challenges faced by the police and the courts everywhere in this arena. But we believe that it is possible to do better. A first step is to document the performance of the police and courts through final disposition, rather than measuring success on the basis of whether the investigation results in an arrest. Given the complexities and long lags in court disposition, it is not feasible to tabulate case dispositions in, say, an annual report. But it would surely be of value to monitor the percentage of arrests that are dismissed by the courts within, say, the first year. That statistic is relevant to monitoring both police and court performance.

What can be done to increase the likelihood that shooters will be arrested and convicted? Lighter caseloads for police investigators and prosecutors would surely help for both fatal and nonfatal shooting cases. But we need more information about the best use of criminal-justice resources to accomplish this important purpose. It is a woefully under-researched area and thus offers an opportunity for scholars interested in policy research. For example, a specific challenge for nonfatal cases, securing the cooperation of victims, could inspire experimentation with a range of behavioral interventions to overcome the widespread hostility to the police and make cooperation more rewarding and less risky. Developing the evidence base in this area is both urgent and feasible.

Acknowledgments

This chapter is based on a report to the Durham, NC, Chief of Police (Cook & Vila 2023). Data from the records of the Durham Police Department were compiled by Jason Schiess, Analytical Services Manager. He also consulted on data processing and interpretation and contributed to the appendix. Additional data were provided by the Office of the Durham District Attorney. Tianyi Zhang and Pilar Kelly assisted with presentations. The Wilson Center for Science and Justice provided financial support for Audrey Vila. Angie Weis Gammell and Brandon Garrett provided helpful comments on an earlier draft. The authors bear sole responsibility for any errors of fact or interpretation.

Notes

1. In FS cases during the five-year period, there have been two trials. A fatal shooting incident in 2018 had a total of six defendants. One was acquitted at trial, and one convicted following a plea to second-degree murder. (The other cases were pending.) Given our definitions, we coded that case as a conviction, since the incident did result in at least one conviction. Another fatal shooting in 2018 resulted in trial and conviction for first degree murder.
2. In fact the total number of individuals and charges is somewhat higher. The table includes all incidents in which there was at least one arrest and one charge. But in cases with two defendants with different outcomes, the incident is classified according to the more successful prosecution outcome, and the second defendant is only included if they had the same outcome. For example, if cases where two defendants were convicted, both are included in the second row, but otherwise the second defendant is dropped from the tabulation.
3. Stirling Whitted was convicted in October 2019, of first degree murder for a fatal shooting on July 31, 2018.
4. An additional five NFS cases had all charges dismissed, but the defendants were prosecuted and convicted for another, still more serious crime.
5. Assuming that the cases that are federally indicted result in conviction.

References

Abrams, David S. (2012) "Estimating the deterrent effect of incarceration using sentencing enhancements" *American Economics Journal: Applied Economics* 4(4), 32–56.Altholz R. (2020) *Living with Impunity: Unsolved Murders in Oakland and the Human Rights Impact on Victims' Family Members*. Berkeley, CA: University Cal. Berkeley Sch. Law. https://www.law.berkeley.edu/wp-content/uploads/2020/01/Living-with-Impunity.pdf

Anderson, E. (1999) *Code of the Street: Decency, Violence, and the Moral Life of the Inner City*. New York: Norton.

Braga, Anthony A. and Philip J. Cook (2018) "The association of firearm caliber with likelihood of death from gunshot injury in criminal assaults" *JAMA Network Open* (released July 27, 2018).
Braga, Anthony A. and Philip J. Cook (2023) *Policing Gun Violence: Strategic Reforms for Controlling Our Most Pressing Crime Problem.* Oxford: Oxford University Press.
Braga, Anthony A. and Desiree Dusseault (2018) "Can homicide detectives improve homicide clearance rates?" *Crime & Delinquency* 64(3): 283–315.
Brunson, Rod K. and Brian A. Wade (2019) "Oh hell no, we don't talk to police: Insights on the lack of cooperation in police investigations of urban gun violence" *Criminology & Public Policy* 18(3): 637–648.
Bureau of Justice Assistance, US Department of Justice (2018) *Promising Strategies for Strengthening Homicide Investigations Findings and Recommendations from the Bureau of Justice Assistance's Homicide Investigations Enhancement Training and Technical Assistance Project* PERF, Wash. DC. https://bja.ojp.gov/sites/g/files/xyckuh186/files/Publications/promising-strategies-for-strengthening-homicide-investigations.pdf
Chalfin, Aaron and Justin McCrary (2018) "Are U.S. cities underpoliced? Theory and evidence" *Review of Economics and Statistics* 100(1): 167–186.
Chalfin, Aaron and Justin McCrary (2017). "Criminal deterrence: A review of the literature" *Journal of Economic Literature* 55(1): 5–48.
Cook, Philip J. (1980) "Reseasrch in criminal deterrence: Laying the groundwork for the second decade" in *Crime and Justice: An Annual Review of Research* (vol 2) eds. Norval Morris and Michael Tonry (Chicago: University of Chicago Press) 211–268.
Cook, Philip J. (2020) "Thinking about gun violence" *Criminology & Public Policy* 19(4): 1371–1393.
Cook, Philip J., Anthony A. Braga, Brian S. Turchan, and Lisa M. Barao (2019) "Why do gun murders have a higher clearance rate than gunshot assaults?" *Criminology & Public Policy* 18(3): 525–551.
Cook, Philip J., Jeffrey Ho, and Sara Shilling (2017) *Criminal Investigations of Gun Assaults and Murders in Durham, 2015: The Challenge of Securing Victim and Witness Cooperation*. Unpublished manuscript, Sanford School of Public Policy, Duke University.
Cook, Philip J. and Javier Lopez (2024) "Explaining the extraordinary decline in Chicago's homicide arrest rates, 1965 to 1994 and beyond: Trends in case mix versus standards for arrest." *Journal of Contemporary Criminal Justice* 40(1): 82–112.
Cook, Philip J. and Jens Ludwig (2022) "Gun violence is THE crime problem" *Vital City* Issue 1, May. https://www.vitalcitynyc.org/issues/issue-1
Cook, Philip J. and Ashley M. Mancik (2024) "The sixty-year trajectory of homicide clearance rates: Toward a better understanding of the great decline" *Annual Review of Criminology* 7: 59–83.
Cook, Philip J. and Audrey Vila (2023) *Gun Violence in Durham, NC, 2017–2021: Investigation and Court Processing of Fatal and Nonfatal Shooting Cases.* Sanford School of Public Policy. https://sanford.duke.edu/sites/default/files/Final%20Draft%20Durham%20Police%20Report.pdf.
Daly, M. (2023) "Inequality, grievances, and the variability in homicide rates" *Evolution and Human Behavior* 44(3): 296–304.

Forst, Brian (2024) "Blackstone's rule has limited our ability to think" *Notre Dame Journal of Law, Ethics & Public* Policy 38 Online supplement, 521–528.

Friedersdof, Conor (2021) "Criminal-justice reformers chose the wrong slogan" *The Atlantic*, August 8.

Johnson, Rucker and Steven Raphael (2012) "How much crime reduction does the marginal prisoner buy?" *Journal of Law and Economics* 55(2): 275–310.

Keel, T. G., J. P. Jarvis, and Y.E. Muirhead (2009) "An exploratory analysis of factors affecting homicide investigations: Examining the dynamics of murder clearance rates" *Homicide Studies* 13: 50–68.

Kegler, S. R., T. R. Simon, M. L. Zwald, et al. (2022) "Vital signs: Changes in firearm homicide and suicide rates — United States, 2019–2020" *MMWR Morb Mortal Wkly Rep* 71: 656–663.

Leovy, Jill (2015) *Ghettoside: A True Story of Murder in America.* New York: Spiegel & Grau.

Mancik, Ashley M. and Karen F. Parker (2019) "Homicide clearances during pre- and post-U.S. crime drop eras: The role of structural predictors and demographic shifts, 1976–2015" *Crime Justice* 42(3): 237–256.

Nagin, Daniel S. (2013) "Deterrence in the twenty-first century" in *Crime and Justice: A Review of Research* (vol. 42) ed. Michael Tonry (Chicago: University of Chcago Press) 199–263.

Ousey, G. C. and M. R. Lee (2008) "Racial disparity in formal social control: an investigation of alternative explanations of arrest rate inequality" *Journal of Research in Crime and Delinquency* 45(3): 322–355.

Phillips, Scott, Dae-Young Kim, Irshad Altheimer, and Greg Drake (2022) "Implementing a stand-alone investigative unit to clear non-fatal shooting cases" *Policing* 16(1): 204–221.

Police Executive Res. Forum (2019) *Review of the Chicago Police Department's Homicide Investigation Process: Observations and Recommendations.* Rep., IAPA, Chicago. https://home.chicagopolice.org/homicideclearancereport2019/

Wellford, Charles, Cythia Lum, Thomas Scott, Hather Vovak, and J. Amber Scherer (2019) "Clearing homicides: Role of organizational, case and investigative dimensions" *Criminology & Public Policy* 18(3): 553–600.

White, Kailey, Philip J. Cook, and Harold A. Pollack (2021) "Gunshot-victim cooperation with police investigations: Results from the Chicago inmate survey *Preventive Medicine* 143.

Zimring, Franklin E. (1968) "Is gun control likely to reduce violent killings?" *University of Chicago Law Review* 35: 721–737.

Zimring, F. E. (1972) "The medium is the message: Firearm caliber as a determinant of death from assault" *The Journal of Legal Studies* 1: 97–123.

Zimring, Franklin E. and Gordon J. Hawkins (1973) *Deterrence: The Legal Threat in Crime Control.* Chicago: University of Chicago Press.

3

Assessing the Impact of Gunshot Detection Technology: Methodologies, Analytical Approaches, and Insights

Daniel S. Lawrence and Eric L. Piza

3.1 Introduction

In the wake of high-profile violent crimes involving firearms, law enforcement and policymakers are renewing their focus on strategies to reduce gun violence. Although prominent mass-shooting events bring much needed attention to the issue, the everyday shooting incidents in local (mostly urban) neighborhoods are largely ignored by the public and media. In 2022, there were 19,651 homicides committed in the United States using a firearm, accounting for 79% of all homicides (CDC, 2024). Much of this gun violence tends to be concentrated within small geographic areas and networks of both victims and perpetrators (Braga, 2003, 2007; Braga et al., 2010; Irvin-Erickson et al., 2017a; Papachristos & Bastomski, 2018). For example, a social network analysis revealed that nearly one-third of the shootings in Newark, NJ, were concentrated within a network comprising less than 4% of the city's total population (Papachristos et al., 2015). These violent firearm-related crimes can have significant impacts on the economic stability and overall quality of life within the communities where they occur. When a community is plagued by firearms violence, property values tend to decrease and business opportunities are lost as businesses close prematurely due to people avoiding the area (Irvin-Erickson et al., 2017b). Individuals who live in communities with gun violence also face diminished quality of life as they must focus on keeping themselves and their families safe, which can lead to the development of adverse health issues such as high blood pressure and anxiety (Cook & Ludwig, 2000).

DOI: 10.1201/9781003430773-4

To respond to firearms violence, many law enforcement agencies across the United States have implemented gunshot detection technology (GDT) in their jurisdictions. A nationally representative survey of law enforcement agencies in 2014 estimated that 16.2% of large agencies (250 or more sworn officers) had used GDT in the past two years, and another 4.0% of large agencies planned to implement the technology in the next two years (Strom et al., 2016). The 2020 LEMAS survey estimated that approximately 2.8% of agencies used a GDT nationally, and this increases to 26.9% among large agencies (DOJ, 2023).

Gunshot detection technology holds the promise of enhancing law enforcement's awareness of firearm-related incidents, thereby enabling targeted police presence in communities plagued by shootings. This heightened police visibility may foster a sense of guardianship among community members and potentially deter criminal activity (Hollis-Peel et al., 2011). Moreover, GDT systems have the capacity to consistently alert officers to shooting incidents, facilitating prompt responses and improving the collection of crucial evidence upon their arrival at the scene. The technology also addresses numerous deficiencies of existing records management systems by automatically assigning officers to shooting events and swiftly delivering crucial information. Traditional methods often involved distressed community members to verbally relay location and other important details to a dispatcher, who then enters this information into the CAD system for officers to access. In contrast, GDT alerts automatically provide precise location data, potentially enabling faster and more direct officer deployment.

Such swift and decisive actions by law enforcement can contribute to the implementation of swift and certain sanctions, a known deterrent to criminal behavior (Durlauf & Nagin, 2011). This is reinforced by the rational choice perspective, which posits that individuals contemplating criminal acts assess the potential gains against the risks involved (Cornish & Clarke, 1986). By promptly reacting to GDT alerts, law enforcement agencies communicate to potential wrongdoers that the likelihood of being caught is heightened, thereby amplifying the perceived risks associated with engaging in criminal behavior (Cornish & Clarke, 1986). This increased perceived risk acts as a deterrent, dissuading would-be offenders from carrying out illicit activities.

However, there remain many questions to the actual impact that GDT may have on policing operations and criminal behavior, particularly firearm shootings. This chapter explores the history and impact of GDT on policing practices aimed at addressing firearm violence. It provides an in-depth exploration of the methodological considerations essential for evaluating the effectiveness of GDT, alongside broader societal implications of its deployment. The chapter reviews previous studies, detailing findings across various outcomes that have been examined. It concludes with a discussion of potential areas for future research and scholarly inquiry in this field.

3.2 History and Types of Gunshot Detection Technology

The earliest efforts to identify the origin of gunshots date back to 1975 when Reid (1975) attempted to track bullets in flight using microphones at the Naval Surface Weapons Center. Two decades later, in response to high levels of firearm-related crimes in Menlo Park, California, Lahr and Fischer (1993) used five wireless microphones to locate high-intensity impulse noise but were unable to distinguish gunshots from other similar sounds. During the early 1990s, Alliant Technology Systems, Inc. began developing the System for Effective Control of Urban Environment Security (SECURES) system to detect gunshots in urban environments (Aguilar, 2015). This early GDT was designed to pinpoint the direction and source of gunfire, providing precise location and directional data. The system was oriented towards potential gunfire locations, enabling users to accurately identify the area in front of them where the gunfire occurred. As such, it was first applied in military settings during the Gulf War to better identify sniper positions. The SECURES system was first installed domestically at Fort Benning's Olympic Village in 1996 during the Olympic Games of Atlanta. Based on its success in Atlanta, the technology was later expanded to locations around the United States (Aguilar, 2015).

In 1995, Trilon Technologies was established and launched its own GDT system known as ShotSpotter, which was implemented in Redwood City in April of 1996 (Mazerolle et al., 2000). In 2009, ShotSpotter acquired the SECURES system, thereby becoming the market-dominating GDT system in the United States (Aguilar, 2015). This system employs a network of acoustic sensors to triangulate the location of gunfire. The operational mechanism involves detecting an acoustic anomaly within the sensor network, wherein at least four sensors collaborate to determine the source of the sound. These data are then transmitted to the vendor for review, where the alert is either confirmed as a gunshot or dismissed as a false positive. Once validated, the alert is forwarded to the agency's computer-aided dispatch system, enabling patrol officers to respond promptly, typically within 60 seconds. ShotSpotter is now one technology under the control of the company SoundThinking.

Other approaches to GDT exist as well. One variant of GDT integrates with closed-circuit television (CCTV) cameras, focusing on monitoring a specific location. This system functions similarly to early military versions but is confined to a defined area, detecting gunfire within a certain radius of the sensor. Upon detecting a gunshot, the system alerts a camera operator, thereby facilitating a quicker investigation than a typical law enforcement response (see Ratcliffe et al., 2019, as an example).

While these are the primary GDT systems in use today, it is important to note the potential for further innovation. For instance, there are concepts for integrating GDT with fire alarm systems in schools to provide real-time alerts to law enforcement during shooting incidents (Onofrio, 2024). Other proposals include linking GDT to drone programs for rapid on-scene assessments (WLKY, 2018) and even embedding the technology in streetlights to enhance urban surveillance capabilities (KOVA Corp, 2021). Although these applications are not yet widespread, they exemplify the expanding horizons of GDT and its potential future applications.

3.3 Limitations and Ethical Considerations

3.3.1 Limitations of the Technology

Numerous studies have consistently affirmed the accuracy of ShotSpotter, the predominant GDT used in the country. Past research has observed ShotSpotter is able to accurately identify 70% of test gunfire in St. Louis (Mares, 2021); 81% in Redwood, CA (Mazerolle et al., 2000; Watkins et al., 2002); 90% in Las Vegas, NV (Koren, 2018); and over 99% in Charleston, SC (Goode, 2012; NIJ, 2007). However, the technology is less accurate in instances where the gunfire is within a closed environment, such as from inside a vehicle, or when close nearby buildings result in the gunfire's echoing also being captured by the technology. Recent media coverage has highlighted that several aspects of GDT accuracy remain underexplored; however, recent efforts to enhance the transparency of GDT systems represent a significant step forward in addressing these concerns (see Feathers, 2024). For example, although ShotSpotter GDT has a fairly reliable system for distinguishing gunshots from fireworks (Aguilar, 2015; Chacon-Rodriguez et al., 2011), many researchers often adopt a more cautious strategy. They exclude events occurring on dates with a higher likelihood of fireworks, such as New Year's Day and Independence Day, to minimize the inclusion of fireworks and excessive celebratory gunfire incidents (Lawrence & Novak, 2024; Piza et al., 2025a).

Community members often lack awareness of how GDT functions and the specific areas it covers within a city. Lawrence et al. (2019a) conducted focus groups to explore community perceptions of police responses to firearm violence, including the use of GDT. Many participants mistakenly believed that the entire city was covered by GDT, leading them to assume they did not need to call 911 to report gunfire in their neighborhoods, as they thought the system would automatically notify authorities. Conversely, this lack of knowledge about the exact locations of GDT sensors could potentially influence crime patterns beyond the covered areas. This may create a "halo effect" or diffusion of benefits (Clarke & Weisburd, 1994), where potential offenders believe they are being monitored even in areas without GDT, thereby potentially deterring criminal activities in those locations as well.

3.3.2 Privacy Concerns

A major concern in the use of GDT is its potential to capture nearby individuals' voices, raising the possibility of police agencies using the technology for persistent or targeted voice surveillance (Policing Project, 2019). Further increasing this concern is the potential ability for agencies to track individuals across a city using the network of acoustic sensors. Even when used for legitimate purposes, such as with sound recordings that are collected around validated gunfire, there is a concern that the recordings could be used in ways not originally approved by city councils, such as presenting recorded voices collected in the audio at the time of the gunshot during investigative and court proceedings. The ultimate concern is that the GDT may be introduced to a city under the guise of identifying gunshots for the police agency but could be exploited for voice surveillance.

While these concerns are valid and warrant vigilance from cities, including the implementation of quality assurance measures to prevent such misuse, a review conducted by a third-party academic nonprofit found that the risk of SoundThinking's GDT being used for voice surveillance is extremely low (Policing Project, 2019). The auditors advised that they reached this conclusion based on their assessment of the safeguards built into how the GDT operates.

Additional recommendations were offered to further safeguard privacy of community members, including shortening the length of recorded audio files, keeping sensor locations hidden from police agencies, and limiting personnel who have access to sensor audio, among others.

3.3.3 Over-Policing

The placement of GDT is typically guided by an empirical analysis of 911 calls reporting shooting events and related crime incident reports. Cities aim to implement GDT in the most efficient and effective way, prioritizing areas with a history of firearm violence. Given that gun violence is closely associated with other violent crimes, GDT is often deployed in areas that already experience a high police presence. Consequently, some critics argue that GDT contributes to over-policing. Moreover, because many communities of color face systemic challenges and reduced opportunities that often result in low socioeconomic status and high crime levels, they are frequently the locations where GDT is implemented. Alerts from GDT can further prompt a variety of law enforcement responses that might not occur without the technology, such as dispatching officers to the alert location, reorienting CCTV cameras to the alert direction, integrating data into predictive policing software, and reinforcing existing stereotypes about certain neighborhoods.

While this topic is in need of further research, a recent evaluation in Chicago found that GDT systems may not generate racial disparities *above and beyond* what results from the standard police response to gunfire (Piza et al., 2025a). Using bivariate Knox and point process tests, the analysis found that both GDT alerts and shots fired calls for service (CFS) occurred in close spatial and temporal proximity to arrests and stops of civilians. Furthermore, the relative effects of GDT and CFS were consistent across civilian race in the majority of instances (Piza et al., 2025a). In light of these findings, Piza et al. (2025a) concluded that any racial disparities resulting from police responses to gunfire likely relate to aspects of the reporting and dispatch processes generally rather than as they relate specifically to GDT.

3.3.4 Ethical Dimensions of Data Science

Data ethics extends beyond the security measures provided by vendors, as researchers may misuse GDT data in ways that could lead to inaccurate findings or harm marginalized communities. This has particular importance given police have become highly reliant on surveillance technologies and the data they generate on a daily basis (Ferguson, 2017; Gaub & Koen, 2021). The promise to identify and communicate gunshots in a more efficient and reliable manner makes GDT a valuable addition to the suite of "big data" sources that are increasingly being leveraged in the social sciences (O'Brien, 2024). Prior research has used GDT alerts to explore research questions unrelated to the programmatic outputs and outcomes of GDT. For example, Mazeika and Uriarte (2019) used GDT alerts to calculate more precise estimates of near repeat shooting patterns than what is possible using only reported crime incidents, and Wheeler et al. (2020) used GDT alerts to test the accuracy of various address-based geocoding tools.

Despite the promise and utility of big data sources generated by police technology, concerns and considerations related to the technology's limitations, potential for voice surveillance, and the likelihood of increased policing in already heavily policed areas raise further issues about the data collected by these systems. This reflects general arguments about the unintended negative consequences of "big data" policing initiatives resulting from the mass deployment of surveillance technologies (Ferguson, 2017). This can be exacerbated by "predictive policing" algorithms that leverage technology and data to empirically rank places, times, and people according to their estimated risk of crime victimization. Enforcement disparities can occur within communities of color, given the simultaneously high levels of disadvantage, crime, and police technology deployment in these areas (Chu et al., 2023; Wheeler, 2020). The emergence of such enforcement disparities can potentially generate biased feedback loops, in which people of color are disproportionately captured in police data and targeted for increased intervention as a result (Ferguson, 2017; Lum & Isaac, 2016).

The application of "big data" sources such as GDT should be guided by an ethical approach guided by transparency of actions and decisions, concern for equality of service, and restraint regarding actions that exacerbate inequality – the hallmarks of democratic policing (Manning, 2010). A data ethics approach requires researchers to focus on the interplay among hardware, software, and data, rather than solely on the diverse digital technologies that facilitate their use (Floridi & Taddeo, 2016). Under a data ethics approach, it is crucial to thoroughly consider the ethical implications from the initiation to the conclusion of a research project. In the context of GDT, this is especially important when using data from a technology that is predominantly deployed in communities of color, regardless of the historical and valid evidence supporting its deployment in these areas.

3.4 Analytical Considerations

3.4.1 Program Operations

A critical component to assessing the impact of GDT is understanding how it has been implemented within policy and practice. Research by Lawrence and Novak (2024) identified varying degrees that GDT programs were implemented across three cities. The authors present a tiered approach of basic to advanced levels of GDT implementation, identified through discussions with stakeholders familiar with GDT operations. They detail Tier 1 programs being those with the most basic response to GDT alerts, where officers merely respond to the scene and perform superficial inspections for gunshot victims, potential suspects, or individuals fleeing the area. This minimal approach is often found in agencies overwhelmed with high-priority criminal activities. Here, officers are required to investigate GDT alerts, but if no individuals in distress are found or there's no immediate evidence that a shooting occurred, the call is marked as unfounded, and the officers move on to the next urgent assignment. This may, in fact, be what's happening in cities that have noted that GDT alerts frequently result in "unfounded" dispositions, such as reviews in Chicago (COIG, 2019), New York City (NYCC, 2024), Philadelphia (Ratcliffe et al., 2019), and Kansas City (Piza et al., 2025b). The increased workload from more gunshot notifications, coupled with an already heavy load of other calls, likely limits officers' ability to engage in more thorough activities that would enhance the program's success. However, more research is necessary to confirm this hypothesis.

Tier 2 programs expand on the basic Tier 1 response by including thorough on-site canvassing after securing the scene. This involves searching for shell casings, firearms, and bullet damage, as well as engaging with residents. Tier 3 programs represent a more comprehensive approach, incorporating advanced technological resources alongside Tier 2 activities. This tier places a strong emphasis on retrieving and analyzing shell casings through the ATF's National Integrated Ballistic Information Network, which can be used to identify suspects in significant crimes by linking them to minor incidents.

Additional tools such as the eTrace program for verifying firearm ownership, surveillance footage analysis, and legal extraction of cellular data further enhance investigative efficiency. Gunshot detection technology systems can also be integrated with public surveillance networks, where cameras can be programmed to focus on the scene of a gunshot identified by the GDT (Lawrence et al., 2024; Ratcliffe et al., 2019; Vovak et al., 2021). Integrating GDT with these auxiliary resources is critical for maximizing investigative efficiency.

Two elements are integral to each of the previously mentioned tiers. The first is the role of a program supervisor, typically at the sergeant level, who oversees program management, coordinates between vendors and detectives, and ensures officer accountability in shooting-related cases. The level of involvement of these supervisors varies, with some ensuring the program's functionality and others actively facilitating forensic analysis and investigating shooting scenes. The second crucial component involves community outreach to foster transparent communication about GDT's operation, coverage, and privacy implications. Addressing these misconceptions is vital for building trust and encouraging proactive incident reporting within the community (Henning et al., 2024).

By knowing how officers respond to GDT alerts, the processes they are expected to do compared to what they actually do, and how an agency manages the program are all critical aspects to better understanding why certain impacts are observed. Such aspects should not be ignored.

3.4.2 Comparisons between Data

A common method for evaluating GDT involves comparing 911 calls related to gunfire to alerts generated by GDT, for example, when assessing accuracy, response times, or dispositions. Ensuring accurate "apples-to-apples" comparisons in such analyses is crucial but often overlooked in existing literature, leading to potential discrepancies and misinterpretations. One key consideration is the type of shooting events being compared. For instance, CFS categorized as "Shots Fired/Heard" are not equivalent to calls for a "Shooting" event, which typically involves an individual being struck by gunfire. Lawrence and Novak (2024) argue that officers are likely to prioritize incidents where they believe someone is injured and in need of lifesaving intervention and respond differently. Therefore, it is essential to focus on CFS that most closely align with the capabilities of GDT, such as "Shots Fired/Heard" calls, to ensure the comparison accurately reflects the scope and effectiveness of the technology.

Second, an issue arises when a single GDT alert is associated with multiple CFS, or vice versa, when one CFS is associated with many GDT alerts. In an ideal scenario, a GDT alert might trigger a dispatch system and be accompanied by a 911 call from a community member who heard the gunshots. In this instance, a one-to-one correspondence exists between the GDT alert and the CFS. However, additional calls from other community members reporting the same incident can lead to a one-to-many ratio, where multiple CFS are counted against a single GDT alert, thus overestimating the calls specific to a single shooting event. Conversely, there are scenarios where multiple GDT alerts correspond to a single CFS. For instance, a community member might report hearing several gunshots from a gunfight moving along a street, leading to multiple GDT alerts for one CFS. Counting these calls and alerts in their raw form would be inappropriate and misleading.

To address this, a straightforward solution involves using geospatial analysis tools such as ArcGIS's "Find Space-Time Matches" to identify and filter out duplicate events within a specific time and geographical frame. While there is no universally perfect frame (see Mares, 2022, for more details), a common practice is to use a window of five minutes and a quarter-mile radius to ensure more accurate comparisons. Lawrence and Novak (2024) adopted this approach in their study examining shooting notifications, identifying 275 overlapping notifications (2.04% of all notifications) that were subsequently removed from analysis. Other scholars may employ broader frames depending on the context of their studies.

3.4.3 Research Methods and Geographical Areas of Examination

When examining the impact of GDT, several challenges and considerations to conduct a rigorous evaluation quickly become apparent. A key consideration when developing statistical models to assess the impact of GDT involves selecting the appropriate geographical areas of examination. Potential areas that outcomes can be measured within can include street segments, census blocks or tracts, specific areas (e.g., coverage areas, police districts, neighborhoods), entire cities, or counties. Generally, more localized areas yield more accurate findings. For instance, Doucette et al. (2021) evaluated the impact of GDT on firearm homicides and arrests using data from 68 urban counties from 1999 to 2016. They found that GDT had no significant impact on firearm-related homicides or arrests. However, they acknowledged a primary limitation: the inability to directly attribute outcomes to the specific areas covered by GDT. It is highly probable that the GDT coverage in the examined counties was relatively small compared to each county overall, most of which likely did not have GDT coverage. City-wide analyses face similar challenges unless the GDT system covers the majority of the city. Thus, using smaller geographical areas can facilitate more accurate assessments of GDT's impact.

One straightforward approach is to compare the change in crime levels in GDT areas before and after deployment. While this method provides a clear and understandable measure of impact, it cannot account for changes occurring in other parts of the city, potentially rendering the assessment of GDT's effectiveness inconclusive. To estimate the impact of an intervention more precisely, researchers will identify a comparison area that is equivalent to the treatment area prior to the deployment of the intervention. For example, including a difference-in-differences (DiD) estimator in statistical models will allow the research to compare the changes in outcomes between a GDT coverage area and an equivalent, non-intervention area. To do this effectively, the researchers must ensure that pre-intervention conditions in the comparison group are equivalent to those in the intervention group, a task that can be quite challenging. Typically, GDT is deployed in areas with high levels of historical crime and firearm incidents, which is a logical approach given the technology's expense and the need to maximize its impact.

Consequently, conducting randomized controlled trials in this context is impractical, if not impossible. Instead, researchers must rely on post-hoc, quasi-experimental methodologies. However, these approaches also present significant challenges, particularly when trying to identify suitable comparison units that do not have GDT but exhibit similar crime characteristics to those areas where GDT is deployed.

Since GDT is often implemented in response to high gun violence, available comparison areas typically have lower crime rates, which complicates the analysis. Additionally, GDT deployments are rarely static; they tend to expand over time, making it harder to maintain consistent comparison groups. For instance, in cities like Chicago, Milwaukee, Richmond, and others, GDT was deployed in a staggered fashion over time, with deployments varying in timing and scale. This uneven deployment complicates the identification of comparable areas, as the remaining sections of the city without GDT – which become fewer and fewer as the GDT coverage area expands – often have lower levels of gun violence. Thus, while quasi-experimental designs offer a feasible alternative to randomized controlled trials, they require careful consideration of the unique deployment patterns and crime dynamics within cities to ensure the validity and reliability of the evaluation. This underscores the complexity of assessing the effectiveness of GDT and highlights the need for sophisticated analytical techniques to address these methodological challenges. For instance, without knowledge of Milwaukee's geographical and temporal expansions of its four GDT coverage areas, a person could mistakenly believe that GDT alerts increased by 407% in the city when comparing the technology's first-year deployment (n = 1,457) to its fifth-year deployment (n = 7,386). But the reality is that the GDT coverage areas expanded from 2.1 to 12.7 square miles during that time, an increase of 505%.

3.4.4 Statistical Models

To assess the impact of GDT, researchers have a variety of statistical models at their disposal. The choice of model depends on several factors, including the unit of analysis, how the outcomes are measured, and the inclusion of comparison units. A careful review of the underlying data, statistical model assumptions, and the control variables to include is essential to ensure the robustness of the applied models. Among the most commonly employed approaches in current research examining GDT are interrupted time series models, panel regression models, entropy balancing, propensity score matching, and various types of synthetic control methods. These methodologies are particularly suited to effectively evaluating the impacts of GDT programs, providing a nuanced understanding of their effectiveness and guiding evidence-based policy decisions.

Time-series data are a widely utilized and powerful tool for conducting impact evaluations, particularly when assessing the effects of GDT on counts of criminal incidents. For example, interrupted time series analysis is an effective method for estimating the impact of an intervention by analyzing a series of data points collected over time. This method involves identifying an intervention point and examining changes in the outcome variable before and after the intervention. Interrupted time series can focus solely on treated areas or include comparison areas through comparative interrupted time series, which estimates the effect of an intervention by comparing the time series of an outcome variable before and after the intervention across treated and untreated groups (Shadish et al., 2002). This approach allows researchers to assess both the immediate impact of an intervention in the time period following its implementation and any subsequent changes in the post-intervention trend relative to the pre-intervention period (Linden, 2015).

Panel regression models, as discussed by Hilbe (2011) and MacDonald and Lattimore (2010), are another robust option for evaluating the impact of GDT. These models account for the structure of panel data, which involves a set of units (e.g., geographic areas) observed over multiple periods. For instance, the unit of analysis might be the "area-month" in a study of GDT, with crime counts that are aggregated monthly for both treatment and comparison areas.

Panel models offer several advantages over methods designed for cross-sectional data or single time series. They provide more accurate and efficient estimates by leveraging the longitudinal nature of the data, allowing for the analysis of complex changes over time and facilitating the control of unobserved heterogeneity (Verbeek, 2008). This capability is particularly beneficial when examining temporal variations in crime rates or the effects of interventions across different regions. When examining count outcomes, such as crime levels, researchers can use Poisson or negative binominal regression models, which offer the advantage of presenting the regression estimates as incidence rate ratios. This allows researchers to more intuitively ascertain the average percentage change in the outcome between the pre- and post-intervention periods (Piza, 2012).

Both interrupted time series and panel count regression models offer sophisticated techniques for analyzing the impact of GDT on criminal incidents, each with unique strengths that make them suitable for different research scenarios. These methods enable a nuanced understanding of how interventions affect trends over time and across different contexts, providing valuable insights for policymakers and law enforcement agencies. However, an effective evaluation of GDT necessitates the inclusion of comparison units. Even when using methods such as interrupted time series and panel count regression models, researchers must identify valid comparison groups before proceeding with the analysis.

There are several matching methods that can be used to identify comparison units and apply appropriate statistical weights where necessary, such as propensity score matching, Euclidean distance matching, and coarsened exact matching, among others. Although recent scholars have argued against the use of propensity scores for matching due to the imbalanced groupings created from its inability to approximate full blocking that's possible with other matching methods (King & Nielsen, 2019), it continues to be a widely employed method for generating comparable groups. Propensity score matching works by using a set of observed covariates to pair comparison units to treatment units based on the conditional probability of their assignment to the treatment group.

It is important to emphasize that for DiD estimation, the data must satisfy the assumption of parallel trends, which requires that the trends of the outcome are similar for both the treatment and comparison groups during the pre-intervention period (Wooldridge, 2010). For instance, if the analysis involves crime and clearance rates, these rates should decline or increase similarly across both groups in the period preceding the intervention. Ensuring common trends is crucial for establishing comparability between the groups and attributing any differences in post-intervention trends to the intervention itself.

There are other methods besides propensity score matching. In a recent study that assessed the impact of GDT on shooting event dispositions (i.e., being classified as unfounded or cleared) and whether ballistic evidence was collected, Piza et al. (2025b) applied entropy balancing methods to create equivalent intervention and comparison groupings. Citing Hainmueller (2012) and Zhao and Percival (2017), they advise that entropy balancing eliminates the need for separate balance testing of the treatment and comparison units by adjusting the weights of covariates into the quasi-experimental function itself based on their propensity for treatment. They argue that this approach is superior to traditional matching methods, such as propensity score matching, when applied to incident-level (as opposed to spatial) data because it does not require iterative model adjustments and balance checks to achieve an adequate balance. Studies have demonstrated that when using incident-level data, entropy balancing outperforms other matching techniques such as coarsened exact matching and propensity score matching (Black et al., 2020; Parish et al., 2018; Zhao & Percival, 2017).

More recently, synthetic control methods have surged in popularity due to their ability to construct a comparative unit from a pool of available, untreated units through weighting. Rather than creating a match to the treatment unit using an actual existing control unit, the synthetic control method generates a synthetic unit using a sampling of comparison units that, once weighted correctly, closely resembles the pre-intervention trends of the intervention area (Saunders et al., 2015). This approach is particularly effective in creating an empirical match for target areas that lack a true counterfactual (Abadie, 2021) or for treatment areas that do not have comparable, untreated control areas within the same jurisdiction (Piza et al., 2020).

Synthetic control methods are robust with respect to covariates and can be rigorously balance-tested to ensure the matching process is appropriately conducted (Ferman, 2021). A variation of this approach, the microsynthetic control method (Robbins et al., 2017), adapts the original synthetic control formula for use with micro-geographic units of analysis, such as street segments.

As previously discussed, a significant challenge in evaluating GDT programs lies in the staggered deployments of the technology. The timing and scale of GDT rollouts are often inconsistent, making it difficult to identify comparable control areas. As the GDT coverage area expands, the number of untreated sections of the city diminishes, further reducing the available areas that exhibit gun violence levels that are equivalent to the treated areas. This disparity complicates the establishment of valid comparison units for evaluation.

Other methodologies have emerged to evaluate program effects in scenarios involving staggered adoption, such as synthetic DiD, staggered synthetic control methods, and the use of group-time averages for DiD models across multiple time periods (Arkhangelsky et al., 2021; Ben-Michael et al., 2021, 2022; Callaway & Sant'Anna, 2021). Each of these approaches offers a unique framework for estimating treatment effects across multiple time periods and treated units. Connealy et al. (2025) recently investigated the effects of GDT in Chicago employing the approach developed by Callaway and Sant'Anna (2021). This method utilizes synthetic control techniques to accommodate multiple intervention units with varying implementation timelines, all within a longitudinal framework.

In conclusion, evaluating the impact of GDT involves a variety of sophisticated statistical models, each tailored to specific research needs and contexts. Researchers must carefully select their models based on the unit of analysis, outcome measures, and the inclusion of comparison units, ensuring robust and accurate results. Common methodologies include interrupted time series, panel regression models, entropy balancing, propensity score matching, and synthetic control methods. Ultimately, the careful application of these diverse methodologies ensures a comprehensive and rigorous evaluation of GDT programs, contributing valuable insights for policymakers and law enforcement agencies in their efforts to reduce gun violence and enhance public safety.

3.5 Effectiveness of Gunshot Detection Technology

Although GDT is often viewed as a state-of-the-art and newly developed technology, research on its applications and effects has been conducted for more than two decades (Mazerolle et al., 1998). Most research on GDT has focused on its procedural and investigative effects, such as the impact on shooting notifications, response times, and evidence collection. The findings in these areas are generally consistent, although with some standouts. Studies typically show that GDT enhances police agencies' awareness of and notifications for shootings (Lawrence & Novak, 2024; Mares & Blackburn, 2021; Mazerolle et al., 1998), shortens officer response times (Choi et al., 2014; Lawrence & Novak, 2024; Mazerolle et al., 1998; Piza et al., 2023), and improves the collection of crime scene evidence, such as shell casings and firearms (CCSVP, 2023; Connealy et al., 2025; Lawrence et al., 2019a; Mares & Blackburn, 2012; Piza et al., 2024a). However, it is crucial to note that GDT also significantly increases officers' workload due to the higher volume of gunshot notifications they must address (Lawrence & Novak, 2024; Blackburn & Mares, 2023). Despite these procedural advantages, research consistently indicates that GDT has minimal, if any, impact on arrest rates and overall crime levels. Multiple studies have found that the technology has little effect on violent crime rates or shooting-related arrests and incidents (Connealy et al., 2025; Lawrence et al., 2019b; Litch & Orrison, 2011; Mares & Blackburn, 2012, 2021; Vovak et al., 2021). Fewer studies have explored other impacts, such as the cost-benefit of the technology (CCSVP, 2023; Lawrence et al., 2019b) and its effects on medical outcomes (Goldenberg et al., 2019). The following section provides a more detailed examination of studies conducted in these domains.

3.5.1 Accuracy

Numerous studies have consistently affirmed the accuracy of ShotSpotter, the predominant GDT used in the county. Past research has observed ShotSpotter able to accurately identify 70% of test gunfire in St. Louis (Mares, 2021); 81% in Redwood, CA (Mazerolle et al., 2000; Watkins et al., 2002); 90% in Las Vegas, NV (Koren, 2018); and over 99% in Charleston, SC (Goode, 2012; NIJ, 2007). Furthermore, other studies have reported that GDT accurately identifies shots fired incidents when compared to CFS (Irvin-Jackson et al., 2017a; Mazerolle et al., 1998, 2000; Renda & Zhang, 2019; Watkins et al., 2002; Wheeler et al., 2020). Different findings have been noted for other types of GDT, such as Philadelphia's non-ShotSpotter GDT that uses a single acoustic sensor connected to a CCTV camera, which Ratcliffe et al. (2019) found to be highly susceptible to false positive events. It is worth noting that the ShotSpotter GDT is less accurate in instances where the gunfire is within a closed environment, such as from inside a vehicle, or when gunfire echoes off nearby buildings or box trucks.

3.5.2 Shooting Notifications

Comparative analysis with 911 calls indicates that GDT alerts provide a more accurate measure of the quantity, timing, and locations of gunfire incidents. For example, Litch and Orrison (2011) found that only 39% of GDT alerts in Hampton, VA, and 43% in Newport News, VA, were associated with 911 calls for confirmed gunfire incidents, indicating that GDT often alerts police to gunfire incidents that would otherwise go unreported. Carr and Doleac (2015, 2016) used GDT data to study reporting rates for homicides and assaults with dangerous weapons in Washington, DC, and Oakland, CA, finding that only 12% of gunshots detected by GDT led to 911 calls reporting the gunfire. In St. Louis, Huebner et al. (2020) examined community calls to police for GDT-detected gunshots and found that residents were more likely to report shootings related to homicides but less likely for other crimes, and communities with a higher proportion of Black residents were less likely to call in shooting events. Piza and colleagues (2023) analyzed geocoded shots-fired incidents reported by both GDT alerts and 911 calls, finding that in over 26% of cases, the reported locations were more than one block apart. This discrepancy indicates police officers responding to gunfire may arrive a significant distance from the actual gunshot location when the more precise geospatial coordinates generated by GDT are unavailable.

The implementation of GDT has significantly increased the volume of shooting notifications that police must respond to. The earliest study on GDT observed an almost fourfold increase (287% increase) in police responses to gunfire following its introduction (Mazerolle et al., 1998). Although more recent studies have not reported such dramatic increases, they still indicate substantial rises in gunshot notifications. For instance, Mares and Blackburn (2021) documented an 80% increase in St. Louis, MO, while Lawrence and Novak (2024) found a 110% increase in Denver, observed increases ranging from 79% to 139% across GDT coverage areas in Milwaukee, and from 65% to 143% in Richmond, CA.

Paradoxically, as police agencies become aware of the dramatic increase in gunfire notifications, they receive decreasing CFS for gunfire incidents. For example, Mares and Blackburn (2012) found that implementing GDT resulted in a drop in shots-fired CFS by community members. In a follow-up analysis, they found CFS were reduced by 30%, but the GDT alerts resulted in an overall increase by 80% of shooting notifications (Mares & Blackburn, 2021).

Mares (2023) employed a longitudinal DiD model to assess the impact that GDT had on two types of firearm-related crimes in Cincinnati, OH: reported gunfire and aggravated assaults with a firearm. The findings revealed that GDT implementation resulted in a 45% decrease in reported gunshots and a 46% reduction in firearm-related aggravated assaults in the targeted census tracts relative to control areas. Piza and colleagues (2024a) found a 22% reduction of shots-fired CFS in the GDT coverage areas. Some of this may result from community members not feeling the need to call the police when they hear gunfire, as they believe the agency will be automatically notified through the GDT, as Lawrence et al. (2019a) observed in discussions with community members during focus groups.

3.5.3 Workload

The implementation of GDT leads to a significant increase in shooting notifications, which in turn has substantially increased the amount of time officers spend responding to these incidents. Lawrence and Novak (2024) reported that officers in Denver spent 2.92 times more staff hours on responding to shooting notifications due to GDT, while in Milwaukee's GDT coverage areas, the increase ranged from 1.78 to 3.36 times more hours. Similarly, in Richmond, officers dedicated between 1.85 and 3.25 times more hours to shooting notifications after the implementation of GDT. In St. Louis, a notable workload increase was observed, with approximately 17,000 GDT alerts adding to the 14,000 shots-fired calls from 2008 to 2018, resulting in an estimated 1,200 additional officer hours annually for responding to these calls (Blackburn & Mares, 2023).

3.5.4 Response Times

When a shooting event is reported, agencies dispatch officers to the scene. Alerts from a GDT are theoretically expected to improve officer response times for several reasons. First, officers have the ability to self-assign to GDT alerts via software on their in-car computers or personal smartphones, allowing them to initiate a response even before the alert is entered into the computer-aided dispatch system. This means officers can begin moving toward the scene ahead of the formal dispatch process. Second, GDT alerts are highly automated compared to traditional CFS, which speeds up officer assignment. Unlike calls that rely on possibly distressed community members to verbally provide location details to a dispatcher, which must then be entered into the CAD system for officers to access, GDT alerts automatically provide precise location data, facilitating quicker and more direct officer deployment.

Previous research provides a mixed perspective on the impact of GDT on police response times. For example, Piza et al. (2024b) found that in Chicago, response times (from dispatch to arrival on scene) to GDT alerts and traditional CFS were nearly identical, differing by just 10 seconds or less. In contrast, an analysis by Topper and Ferrazares (2023) indicated that the implementation of GDT in a Chicago district actually resulted in officers being dispatched one minute later and arriving on scene nearly two minutes later compared to traditional calls. Some studies, however, have reported more favorable outcomes. For instance, Mazerolle et al. (1998) noted a 16% reduction in response times to GDT activations, Choi et al. (2014) observed a 32-second improvement, and Piza et al. (2023) reported GDT alerts occurred a median of 93 seconds prior to the first 911 CFS reporting the same event. An earlier study of GDT in St. Louis by Mares and Blackburn (2012) showed enhanced response times to shots-fired calls; however, more recent study by the same authors (Mares & Blackburn, 2021) found no significant difference in response times following GDT implementation.

Lawrence and Novak (2024) found that, on average, officers arrived at the scene approximately 52 seconds faster in response to GDT alerts than to shots-fired calls across the eight areas studied in Denver, Milwaukee, and Richmond. To gain a more nuanced understanding of these differences, they analyzed response times in distinct segments: (1) the time from notification of a shooting event to the officer assignment, (2) the time from officer assignment to their arrival at the scene, and (3) the overall time from notification of a shooting event to officer arrival. Their findings indicate that the quicker response times for GDT alerts are not necessarily due to officers responding faster to these alerts over shots-fired calls. Rather, the significant advantage of GDT appears to be in its ability to expedite the process from the occurrence of a shooting to the assignment of an officer.

Taking response times a step further, Sandau and colleagues (2024) recently conducted an in-depth analysis of how police and EMS differ in their response times when transporting gunshot victims to hospitals. Their study, which examined 267 gunshot victims in Camden, NJ, assessed the variations in transport trends and times between EMS and police transports for shooting events both initiated by CFS and GDT alerts. The results revealed that response times improved for both EMS and police following the implementation of GDT. Notably, the presence of GDT was linked to a higher frequency of police transports for gunshot victims, highlighting the technology's significant impact on emergency response practices.

3.5.5 Crime Rates

A primary concern in the GDT research relates to the impact of the technology on crime in target areas. While the studies vary in terms of target area operationalization, study period length, and quasi-experimental design, they are generally consistent in finding that GDT has little effect on violent crime occurrence. Two noteworthy exceptions were observed in Cincinnati, OH (Mares, 2023) and Winston-Salem, NC (CCSVP, 2023), where gun assaults were significantly reduced by 45% and overall violence reduced by 24%, respectively, in GDT target areas compared to controls.

Two studies tested the effect of St. Louis's GDT system (Mares & Blackburn, 2012, 2021). Mares and Blackburn (2012) analyzed the effect of the original system deployment, incorporating a 34-month (January 2006–October 2009) student period. While the analysis found a significant reduction of shots fired CFS in GDT target areas, no significant changes were observed for part 1 crime. A follow-up evaluation by Mares and Blackburn (2021) analyzed the effect of GDT across three temporal phases: the initial GDT implementation in 2008, the expansion of GDT coverage in 2013, and four months in 2016. Similar to the original evaluation, Mares and Blackburn (2021) found a reduction of shots fired CFS with no significant changes for violent crime incidents.

Lawrence et al. (2019b) matched GDT target areas with similarly structured control areas using propensity score matching to assess the impact of GDT in Denver, CO; Milwaukee, WI; and Richmond, VA. The implementation of GDT was linked to an uptick in CFS related to shooting incidents across all three cities and an increase in firearm-related crimes in Denver. In Richmond, there was some evidence suggesting a reduction in firearm-related crime, although Milwaukee showed no significant changes in this regard. The study highlighted the dual-directional impact of GDT on shots-fired CFS and provided inconclusive evidence regarding its effects on the measured gun violence outcomes.

Vovak et al. (2021) evaluated the effects of a Wilmington, DE system that combined GDT and CCTV cameras together, focusing on homicide and non-fatal shooting rates. The analysis used Bayesian statistical models to compare Wilmington with similar cities, matched by proximity, population, socio-demographics, and crime rates, but it is noteworthy that only city-level effects were examined. Results found no significant impact on overall crime levels, with both homicides and shootings increasing during the post-intervention period.

Studies incorporating matched case control designs that build upon the strengths of traditional quasi-experimental methods have come to similar conclusions. Piza et al. (2024a) incorporated the microsynthetic control method to test the crime prevention effect of GDT in Kansas City, MO. The street segments falling within the 3.5 square-mile target area were compared to a weighted control group of street segments without GDT. While shots fired CFS were 22% lower in the GDT area post-intervention than the weighted control area during the same period, GDT did not influence any of the gun violence categories involving confirmed victims (non-fatal shootings, fatal shootings, and aggravated assaults and robberies committed with a firearm). Connealy et al. (2025) incorporated a staggered synthetic control design to test the effect of GDT in Chicago, IL, with the system expanding from three square miles to approximately 136 square miles over 11 deployment phases between 2012 and 2018. The GDT had no effect on fatal shootings, non-fatal shootings, general part 1 gun crimes, or shots fired CFS.

Some GDT studies have suffered from issues with spatial delineation, specifically in terms of operationalizing target areas. Litch and Orrison (2011) tested the effect of GDT at the police district level, finding no effect on crime. However, the GDT target areas did not align to police district boundaries in Hampton or Newport News (see Litch & Orrison, 2011, figure 3.2 and figure 3.10), meaning GDT coverage was overestimated. Similarly, Doucette et al. (2021) found no association between GDT and firearms homicides in 68 urban counties from 1999 to 2016. However, given that the GDT system typically covers small proportions of municipalities, the precise GDT coverage was not operationalized, which may have introduced bias into the effect estimates.

Gunshot detection technology's lack of impact on violent crime outcomes stands in stark contrast with non-technological interventions, as strategies such as focused deterrence (Braga et al., 2018), hot spot policing (Braga et al., 2019; Braga & Weisburd, 2022), and problem-oriented policing (Hinkle et al., 2020) have each been associated with significant crime reduction effects. Interestingly, it is more challenging to compare GDT with other police technologies that it is typically integrated with, increasingly within real-time crime center frameworks. While a rich body of literature shows video surveillance cameras have a modest yet significant effect on crime (Piza et al., 2019), the crime control effect of license plate readers has been tested in only three studies (Koper et al., 2013; Koper & Lum, 2019; Lum et al., 2011) and the influence of NIBIN programs on gun crime investigations has predominately been measured through post hoc comparisons of cleared and uncleared cases rather than intentionally designed case control evaluations (Baskin & Sommers, 2010; Flippin et al., 2022; McEwen & Regoeczi, 2015).

3.5.6 Evidence Collection and Clearances

While most GDT outcome evaluations have focused on the crime prevention effect of GDT, a literature on the technology's investigative benefits has begun to develop. This body of research has typically focused on the on-scene collection of physical evidence. Mares and Blackburn (2012) found that GDT alerts in St. Louis, MO, led to ballistic evidence, such as bullet shell casings, being recovered in 2% of cases, as compared to 17% for shots fired CFS in the entire city. The Center for Crime Science and Violence Prevention (CCSVP, 2023) reported that firearms and shell casings in Winston-Salem, NC, were recovered from 3% and 37.1% of GDT alerts, respectively. The recovery of shell casings is especially significant, as the 1,577 casings found within the three-square-mile GDT target area represent roughly 43% of the total 3,678 shell casings recovered in the 134-square-mile city. Lawrence and colleagues (2019a) found a marginally significant increase in the collection of shell casings at shooting scenes in GDT areas in aggregated analyses across three cities, with this increase reaching statistical significance in Richmond. The two recent synthetic control studies found the collection of ballistic evidence and recovery of firearms in Kansas City, MO (Piza et al., 2024a), and the recovery of firearms in Chicago, IL (Connealy et al., 2025), were significantly higher in GDT target areas as compared to the weighted control area.

Research focused on the effect of GDT on case clearance – the percentage of cases solved, most typically through the arrest of a suspect – appears more sparsely in the literature. This is a key knowledge gap, as clearance rates have long been considered a key measure of police performance (Baughman, 2020) and the public prioritizes justice being delivered to victims of crime (Cook & Mancik, 2024). The studies that have focused on this topic find little evidence that GDT increases clearance rates. Lawrence et al. (2019b) found no evidence that GDT impacted the number of cases resulting in arrest in any of the study settings with similar null finding found in Dallas, TX (Mazerolle et al., 1998), and Brockton, MA (Choi et al., 2014). Vovak et al. (2021) found that the GDT system in Wilmington, DE, that was integrated with a pre-existing CCTV system was associated with a decrease in homicide and non-fatal shooting clearance during the post-implementation period. Both Doucette et al. (2021) and Litch and Orrison (2011) found no effect of GDT on case clearance, though the aforementioned limitations with GDT coverage operationalization are worth repeating.

Piza and colleagues (2025b) tested the effect of GDT on investigative outcomes in Kansas City, MO. This analysis considered cases that were cleared either through an on-scene arrest or resulting from a retroactive investigation. Piza et al. (2025b) used the entropy balancing method to create a weighted control group of non-GDT cases to shooting incidents occurring in the GDT target area in Kansas City, MO. Results of the logistic regression models found that GDT did not influence the likelihood of case clearance for either fatal or non-fatal shootings.

3.5.7 Costs

A few studies have explored the cost-benefit relationship of GDT, revealing significant variation in findings. Gunshot detection technology systems, such as ShotSpotter, carry substantial costs, averaging around $65,000 per square mile annually, translating into considerable expenses for metropolitan areas. Evaluating these costs against potential crime reductions facilitated by the technology, alongside the expenses linked to those crimes, allows researchers to gauge the cost-effectiveness of GDT initiatives.

Following the implementation of GDT, Winston-Salem recorded a 24% reduction in violent crime (CCSVP, 2023), leading researchers to conclude that the benefits outweighed the costs. Their estimates suggest that GDT potentially saved the city between $5 and $8 million annually (CCSVP, 2023), indicating a return on investment ranging from $15 to $25 for every dollar spent on the technology.

Lawrence et al. (2019b) conducted analyses in Denver, Milwaukee, and Richmond, yielding varied findings regarding the cost-benefit of GDT. Their study considered societal costs, encompassing expenses related to the criminal justice system, crime victims, pain and suffering, and other jurisdictional costs, both including and excluding the costs associated with homicides.

This distinction was crucial due to the exceptionally high cost of a single homicide, estimated around $10 million when factoring in the victim's lifetime earnings (McCollister et al., 2010).

In Denver, the results indicated substantial societal-level savings. When including homicides, there was a net savings of $1.6 million, whereas excluding homicides showed a smaller net savings of $191,000. Milwaukee, on the other hand, showed a cost of $11.7 million when including homicides, which turned into a savings of $1.5 million when excluding homicides. Conversely, Richmond demonstrated a $13.9 million savings when homicides were included but a cost of $1.4 million when excluding them.

Approaching these cost-benefit analyses requires careful consideration, particularly when incorporating homicides into the estimates. Given that a significant number of homicides involve shootings, and considering that the estimated lifetime, familial, and societal costs of a single homicide can amount to around $10 million, including these high-cost incidents can disproportionately influence overall cost-benefit calculations. This reflects prior arguments that common methodologies can bias crime cost estimates upwards, which provides an unclear picture of fiscal savings for policymakers (Ratcliffe, 2015; Tonry, 2015). Additionally, as Cook and Soliman (2024) argue in their decision not to conduct a cost-benefit analysis in Durham, NC, it is crucial to account for the costs associated with the increased police response when evaluating the financial implications of GDT. Therefore, while GDT may provide some procedural and community safety benefits, justifying its deployment financially demands meticulous and context-specific assessment.

3.5.8 Medical

Few studies have investigated the impact of GDT on medical outcomes following shooting incidents. In the context of police response to gunshot victims, the practice of rapid police transport to trauma centers has highlighted the critical role of first-responding officers in prehospital care. Previous research indicates that these rapid police transports can yield outcomes comparable to (Wandling et al., 2016) or even better than (Brand et al., 2014) those of traditional EMS responses. Specifically focusing on police responses initiated by GDT alerts, Goldenberg and colleagues (2019) analyzed the transport practices of 627 shooting victims in Camden, NJ, comparing cases involving GDT alerts to those initiated by CFS. Their findings revealed no significant difference in mortality rates between the two groups; however, victims in incidents triggered by GDT alerts tended to have more severe injuries, were more likely to be transported by police, less likely to receive trauma bay resuscitative measures, and more likely to require blood transfusions.

3.6 Future Scholarship

3.6.1 Policy and Policy Compliance

The application of GDT as a tool for violence prevention is still relatively new, and the research on its impact remains limited (Strom et al., 2016). There are numerous study areas that warrant further investigation. One key reason for the varying impacts of GDT across different locations and studies is likely tied to the diverse policies that agencies adopt in relation to the technology. Lawrence and Novak (2024) highlight this with their tiered approach to GDT programs; however, their study, while insightful, primarily reflects the authors' perspectives based on their examination of GDT programs in three cities rather than extensive empirical analysis. There is a pressing need for more research to understand how different agencies implement and operate GDT programs. This includes exploring what policing practices and other technologies are integrated or used alongside GDT.

Understanding officer compliance with GDT policies within these various tiers is also crucial, as there may be significant variations. Further research should investigate how GDT policies are communicated to officers, how these policies are received, and what obstacles may prevent officers from adhering to them. A noteworthy feature of GDT is that officers can self-assign to alerts through the system's software before being officially dispatched. Future research should examine how these self-assignments impact response times and adherence to policy guidelines.

3.6.2 Program Cancellations

While many cities have adopted GDT, some have determined that the benefits do not justify the costs, leading them to terminate their GDT contracts. For instance, the Charlotte-Mecklenburg Police Department ended its contract with ShotSpotter in 2016, noting that although the sensors functioned correctly, they did not yield sufficient results to warrant the expense (Vandergriff, 2016). Campaign Zero's website, cancelshotspotter.com, actively encourages the public and agencies to discontinue ShotSpotter contracts. Their research highlights approximately 20 agencies that initially implemented ShotSpotter GDT in their cities but later canceled their contracts. This trend offers researchers a valuable opportunity to study natural experiments and assess the effects of GDT removal on gun-related crime rates.

3.6.3 Police–Community Relationships

Some studies have explored community perceptions regarding the police's use of GDT to address firearm violence. For instance, Lawrence et al. (2019a) conducted focus groups with residents in GDT coverage areas of Milwaukee,

Denver, and Richmond, revealing that although community members exhibited low levels of trust and confidence in the police, they generally supported the idea of legitimate policing to prevent gun crime and catch shooting suspects.

Henning et al. (2024) surveyed residents across 23 neighborhoods in Portland, Oregon, and discovered a range of opinions about the technology. Their findings indicated that residents familiar with the technology and those with privacy concerns were more likely to oppose GDT, while those concerned about gun violence and those with higher trust in local law enforcement tended to support it. Despite these insights, there is a notable gap in research specifically comparing community attitudes toward police responses to shootings detected by GDT versus those reported through CFS, both on a case-by-case basis and in broader contexts.

3.6.4 Other Needed Research

There are several additional areas of research that could deepen our understanding of GDT's usage and impact. While the potential influence on medical outcomes has been mentioned, only a single study has specifically investigated this to date (Goldenberg et al., 2019). More comprehensive research is needed to clarify how GDT assists officers in reaching scenes where timely intervention is critical for saving community members' lives following a shooting.

Previous studies have highlighted significant increases in patrol officers' workload following the introduction of GDT (Blackburn & Mares, 2023; Lawrence & Novak, 2024). Yet, little research has explored how this heightened response to shooting incidents may impact officers' capacity to prevent or respond to crime in other areas of a city. Further investigation is needed to understand the potential broader implications of GDT on policing practices.

Lastly, police agencies are not the only entities within the criminal justice system that utilize GDT data. Detailed forensic reports and audio files from GDT vendors are often reviewed and introduced in court proceedings, yet there is a significant gap in understanding precisely how these materials are used, whether District Attorneys or prosecutors find them beneficial, and the influence such evidence has on jury deliberations. Future research should focus on examining the role of GDT data in court settings and its impact on case outcomes.

3.7 Conclusion

Gunshot detection technology has the potential to significantly enhance law enforcement's ability to identify and respond to firearm-related incidents, facilitating a more strategic police presence in communities affected by gun violence. Although current studies have indicated that GDT has minimal impact on reducing violent crimes and firearm-related incidents, the research supports GDT's role in improving procedural and investigative practices. It is crucial for researchers to conduct their analyses ethically and employ rigorous methodologies and statistical techniques to accurately assess the impact of GDT. As the field advances, it is essential to incorporate best practices and build on previous research to provide law enforcement and the public with a clear understanding of the true effectiveness of this technology.

References

Abadie, A. (2021). Using synthetic controls: Feasibility, data requirements, and methodological aspects. *Journal of Economic Literature, 59*(2), 391–425. https://doi.org/10.1257/jel.20191450

Aguilar, J. (2015). Gunshot detection systems in civilian law enforcement. *Journal of the Audio Engineering Society, 63*(4), 280–291. https://aes2.org/publications/elibrary-page/?id=17585

Arkhangelsky, D., Athey, S., Hirshberg, D. A., Imbens, G. W., & Wager, S. (2021). Synthetic difference-in-differences. *American Economic Review, 111*(12), 4088–4118. https://doi.org/10.1257/aer.20190159

Baskin, D., & Sommers, I. (2010). The influence of forensic evidence on the case outcomes of homicide incidents. *Journal of Criminal Justice, 38*(6), 1141–1149. https://doi.org/10.1016/j.jcrimjus.2010.09.002

Baughman, S. (2020). How effective are police? The problem of clearance rates and criminal accountability. *Alabama Law Review, 72*(1), 47–130.

Ben-Michael, E., Feller, A., & Rothstein, J. (2021). The augmented synthetic control method. *Journal of the American Statistical Association, 116*(536), 1789–1803. https://doi.org/10.1080/01621459.2021.1929245

Ben-Michael, E., Feller, A., Rothstein, J. (2022). Synthetic controls with staggered adoption. *Journal of the Royal Statistical Society Series B: Statistical Methodology, 84*(2), 351–381. https://doi.org/10.1111/rssb.12448

Black, B. S., Lalkiya, P., & Lerner, J. Y. (2020). The trouble with coarsened exact matching. *SSRN Electronic Journal*. https://doi.org/10.2139/ssrn.3694749

Blackburn, E., & Mares, D. (2023). *The hidden costs of police technology: Evaluating acoustic gunshot detection systems*. Police Chief Online. https://www.policechiefmagazine.org/the-hidden-costs-of-police-technology/

Braga, A. A. (2003). Serious youth gun offenders and the epidemic of youth violence in Boston. *Journal of Quantitative Criminology, 19*(1), 33–54.

Braga, A. A. (2007). *Gun violence among serious young offenders*. Department of Justice.

Braga, A. A., Papachristos, A. V., & Hureau, D. M. (2010). The concentration and stability of gun violence at micro places in Boston, 1980–2008. *Journal of Quantitative Criminology, 26*, 33–53. https://doi.org/10.1007/s10940-009-9082-x

Braga, A. A., Turchan, B. S., Papachristos, A. V., & Hureau, D. M. (2019). Hot spots policing and crime reduction: An update of an ongoing systematic review and meta-analysis. *Journal of Experimental Criminology, 15*(3), 289–311. https://doi.org/10.1007/s11292-019-09372-3

Braga, A. A., & Weisburd, D. L. (2022). Does hot spots policing have meaningful impacts on crime? Findings from an alternative approach to estimating effect sizes from place-based program evaluations. *Journal of Quantitative Criminology, 38*(1), 1–22. https://doi.org/10.1007/s10940-020-09481-7

Braga, A. A., Weisburd, D., & Turchan, B. (2018). Focused deterrence strategies and crime control: An updated systematic review and meta-analysis of the empirical evidence. *Criminology & Public Policy, 17*(1), 205–250. https://doi.org/10.1111/1745-9133.12353

Brand, R. A., Salhi, R. A., Holena, D. N., Poell, E., Branas, C. C., & Carr, B. G. (2014). Severity adjusted mortality in trauma patients transported by police. *Annals of Emergency Medicine, 63*(5), 608–614. https://doi.org/10.1016/j.annemergmed.2013.11.008

Callaway, B., & Sant'Anna, P. H. C. (2021). Difference-in-differences with multiple time periods. *Journal of Econometrics, 225*(2), 200–230. https://doi.org/10.1016/j.jeconom.2020.12.001

Carr, J. B., & Doleac, J. L. (2015). *Keep the kids inside?: Juvenile curfews and urban gun violence.* University of Virginia.

Carr, J. B., & Doleac, J. L. (2016). *The geography, incidence, and underreporting of gun violence: New evidence using ShotSpotter data*. Brookings Institution.

Center for Crime Science and Violence Prevention (CCSVP). (2023). *A cost-benefit analysis of ShotSpotter in Winston-Salem, NC: Improving the police response to gunfire.* Southern Illinois University Edwardsville. https://www.siue.edu/ccsvp/pdf/ShotSpotterpublic.pdf

Center for Disease Control and Prevention (CDC). (2024). *National Center for Health Statistics – Assault or homicide.* https://www.cdc.gov/nchs/fastats/homicide.htm

Chacon-Rodriguez, A., Julian, P., Castro, L., Alvarado, P., & Hernandez, N. (2011). Evaluation of gunshot detection algorithms. *Journal of IEEE Transactions on Circuits and Systems, 58*(2), 363–373.

Choi, K., Librett, M., & Collins, T. J. (2014). An empirical evaluation: Gunshot detection system and its effectiveness on police practices. *Police Practice and Research, 15*(1), 48–61. https://doi.org/10.1080/15614263.2013.800671

Chu, S. P., Pezzella, F. S., & Evans, J. D. (2023). Surveillance load – A burden of search borne by Black and Brown bodies. *Critical Criminology, 31*(2), 451–466. https://doi.org/10.1007/s10612-022-09673-6

City of Chicago Office of Inspector General (COIG). (2019). *The Chicago Police Department's use of ShotSpotter technology.* COIG. https://igchicago.org/wp-content/uploads/2021/08/Chicago-Police-Departments-Use-of-ShotSpotter-Technology.pdf

Clarke, R. V., & Weisburd, D. (1994). Diffusion of crime control benefits: Observations on the reverse of displacement. In Ronald V. Clarke (Ed.), *Crime prevention studies, Vol 3.* Criminal Justice Press.

Connealy, N., Piza, E., Arietti, R., Mohler, G., & Carter, J. (2025). Staggered Deployment of Gunshot Detection Technology in Chicago, IL: A Matched Quasi-Experiment of Gun Violence Outcomes. *Journal of Experimental Criminology, 21*(3), 1027–1053. https://doi.org/10.1007/s11292-024-09617-w

Cook, P. J., & Ludwig, J. (2000). *Gun violence: The real costs.* Oxford University Press.

Cook, P. J., & Mancik, A. (2024). The sixty-year trajectory of homicide clearance rates: Toward a better understanding of the great decline. *Annual Review of Criminology, 7*(1). https://doi.org/10.1146/annurev-criminol-022422-122744

Cook, P., & Soliman, A. (2024). *Evaluation of Durham's ShotSpotter installation: Results of a 12-month pilot project.* Wilson Center for Science and Justice at Duke Law. https://wcsj.law.duke.edu/news/durhamshotspotter/

Cornish, D. B., & Clarke, R. V. (1986). *The reasoning criminal: Rational choice perspectives on offending.* Springer.

Department of Justice (DOJ). (2023). *Law Enforcement Management and Administrative Statistics (LEMAS), 2020.* Office of Justice Programs, Bureau of Justice Statistics.

Doucette, M. L., Green, C., Necci Dineen, J., Shapiro, D., & Raissian, K. M. (2021). Impact of ShotSpotter technology on firearm homicides and arrests among large metropolitan counties: A longitudinal analysis, 1999–2016. *Journal of Urban Health, 98*(5), 609–621. https://doi.org/10.1007/s11524-021-00515-4

Durlauf, S. N., & Nagin, D. S. (2011). Imprisonment and crime: Can both be reduced? *Criminology & Public Policy, 10*(1), 13–54. https://doi.org/10.1111/j.1745-9133.2010.00680.x

Feathers, T. (2024, June 25). The mystery of AI gunshot-detection accuracy is finally unraveling. *Wired Magazine.*

Ferguson, A. G. (2017). *The rise of big data policing: Surveillance, race, and the future of law enforcement.* New York University Press.

Ferman, B. (2021). On the properties of the synthetic control estimator with many periods and many controls. *Journal of the American Statistical Association, 116*(536), 1764–1772. https://doi.org/10.1080/01621459.2021.1965613

Flippin, M. R., Katz, C. M., & King, W. R. (2022). Examining the impact of a crime gun intelligence center. *Journal of Forensic Sciences, 67*(2), 543–549. https://doi.org/10.1111/1556-4029.14952

Floridi, L., & Taddeo, M. (2016). What is data ethics? *Philosophical Transactions of the Royal Society A, 374,* 20160360. https://doi.org/10.1098/rsta.2016.0360

Gaub, J. E., & Koen, M. C. (2021). Cameras and police dataveillance: A new era in policing. In B. A. Arrigo & B. Sellers (Eds.), *The pre-crime society: Crime, culture and control in the ultramodern age.* Bristol University Press.

Goldenberg, A., Rattigan, D., Dalton, M., Gaughan, J., Thomson, J. S., Remick, K., Butts, C., & Hazelton, J. P. (2019). Use of ShotSpotter detection technology decreases prehospital time for patients sustaining gunshot wounds. *The Journal of Trauma and Acute Care Surgery, 87*(6), 1253–1259. https://doi.org/10.1097/TA.0000000000002483

Goode, E. (2012, May 28). Shots fired, pinpointed and argued over. *The New York Times.* https://www.nytimes.com/2012/05/29/us/shots-heard-pinpointed-and-argued-over.html

Hainmueller, J. (2012). Entropy balancing for causal effects: A multivariate reweighting method to produce balanced samples in observational studies. *Political Analysis, 20*(1), 25–46. https://doi.org/10.1093/pan/mpr025

Henning, K., Carmony, J., & Peterson, C. (2024). Public support for gunshot detection technology. *Security Journal.* https://doi.org/10.1057/s41284-024-00434-4

Hilbe, J. M. (2011). *Negative binomial regression.* Cambridge University Press.

Hinkle, J. C., Weisburd, D., Telep, C. W., & Petersen, K. (2020). Problem-oriented policing for reducing crime and disorder: An updated systematic review and meta-analysis. *Campbell Systematic Reviews, 16*(2), e1089. https://doi.org/10.1002/cl2.1089

Hollis-Peel, M. E., Reynald, D. M., van Bavel, M., Elffers, H., & Welsh, B. C. (2011). Guardianship for crime prevention: A critical review of the literature. *Crime, Law Social Change, 56*, 53–70. https://doi.org/10.1007/s10611-011-9309-2

Huebner, B. M., Lentz, T. S., & Schafer, J. A. (2020). Heard shots – Call the police? An examination of citizen responses to gunfire. *Justice Quarterly, 39*(4), 673–696. https://doi.org/10.1080/07418825.2020.1799063

Irvin-Erickson, Y., La Vigne, N. G., Levine, N., Tiry, E., & Bieler, S. (2017a). What does gunshot detection technology tell us about gun violence? *Applied Geography, 86*, 262–273. https://doi.org/10.1016/j.apgeog.2017.06.013

Irvin-Erickson, Y., Lynch, M., Gurvis, A., Mohr, E., & Bai, B. (2017b). *Gun violence affects the economic health of communities*. Urban Institute. https://www.urban.org/sites/default/files/publication/90666/eigv_brief_0.pdf

King, G., & Nielsen, R. (2019). Why propensity scores should not be used for matching. *Political Analysis, 27*, 435–454. https://doi.org/10.1017/pan.2019.11

Koper, C. S., & Lum, C. (2019). The impacts of large-scale license plate reader deployment on criminal investigations. *Police Quarterly, 22*(3), 305–329. https://doi.org/10.1177/1098611119828039

Koper, C. S., Taylor, B. G., & Woods, D. J. (2013). A randomized test of initial and residual deterrence from directed patrols and use of license plate readers at crime hot spots. *Journal of Experimental Criminology, 9*(2), 213–244. https://doi.org/10.1007/s11292-012-9170-z

Koren, D. (2018). *ShotSpotter pilot assessment*. Las Vegas Police Metropolitan Police Department.

KOVA Corp. (2021). Smart, safe cities: GE's smart streetlights to include gunshot detection. *KOVA Corp*. https://www.kovacorp.com/smart-safe-cities-ges-smart-streetlights-to- include-gunshot-detection

Lahr, J., & Fischer, F. (1993). *Location of acoustic sources using seismological techniques and software*. U.S. Geological Survey.

Lawrence, D. S., La Vigne, N. G., Goff, M., & Thompson, P. S. (2019a). Lessons learned implementing gunshot detection technology: Results of a process evaluation in three major cities. *Justice Evaluation Journal, 1*(2), 109–129. https://doi.org/10.1080/24751979.2018.1548254

Lawrence, D. S., La Vigne, N. G., & Thompson, P. S. (2019b). *Evaluation of gunshot detection technology to aid in the reduction of firearms violence*. Office of Justice Programs, National Institute of Justice. https://www.ojp.gov/pdffiles1/nij/grants/254283.pdf

Lawrence, D. S., & Novak, K. J. (2024). Firearm shootings and the police response: Examining the impact of gunshot detection technology. *Police Practice and Research, 26*(6), 726–745. https://doi.org/10.1080/15614263.2024.2384879

Lawrence, D. S., Peterson, B. E., Shukla, R., & Robin, L. (2024). *Optimizing the use of video technology to improve criminal justice outcomes*. Office of Justice Programs. https://www.ojp.gov/pdffiles1/nij/grants/308635.pdf

Linden, A. (2015). Conducting interrupted time-series analysis for single- and multiple-group comparisons. *The Stata Journal, 15*(2), 480–500. https://doi.org/10.1177/1536867X1501500208

Litch, M., & Orrison, G. A. (2011). *Draft technical report for SECURES demonstration in Hampton and Newport News, Virginia*. National Institute of Justice. https://nij.ojp.gov/library/publications/draft-technical-report-secures-demonstration- hampton-and-newport-news-virginia

Lum, C., Hibdon, J., Cave, B., Koper, C. S., & Merola, L. (2011). License plate reader (LPR) police patrols in crime hot spots: An experimental evaluation in two adjacent jurisdictions. *Journal of Experimental Criminology, 7*(4), 321–345. https://doi.org/10.1007/s11292-011-9133-9

Lum, K., & Isaac, W. (2016). To predict and serve? *Significance, 13*(5), 14–19. https://doi.org/10.1111/j.1740-9713.2016.00960.x

MacDonald, J. M., & Lattimore, P. K. (2010). Count models in criminology. In A. R. Piquero & D. Weisburd (Eds.), *Handbook of quantitative criminology* (pp. 683–698). Springer.

Manning, P. K. (2010). *Democratic policing in a changing world*. Routledge.

Mares, D. (2021). *False negatives in acoustic gunfire detection: A comparison of gunfire events and calls for service*. St. Louis Metropolitan Police Department.

Mares, D. (2022). *Gunshot detection: Reducing gunfire through acoustic technology* (Problem- Oriented Guides for Police. Response Guide No. 14.). Officer of Justice Programs, Bureau of Justice Assistance.

Mares, D. (2023). Evaluating an acoustic gunshot detection system in Cincinnati. In E. Groff & C. Haberman (Eds.), *Understanding crime and place: A methods handbook* (pp. 396–406). Temple University Press.

Mares, D., & Blackburn, E. (2012). Evaluating the effectiveness of an acoustic gunshot location system in St. Louis, MO. *Policing: A Journal of Policy and Practice, 6*(1), 26–42. https://doi.org/10.1093/police/par056

Mares, D., & Blackburn, E. (2021). Acoustic gunshot detection systems: A quasi-experimental evaluation in St. Louis, MO. *Journal of Experimental Criminology, 17*, 193–215. https://doi.org/10.1007/s11292-019-09405-x

Mazeika, D. M., & Uriarte, L. (2019). The near repeats of gun violence using acoustic triangulation data. *Security Journal, 32*(4), 369–389. https://doi.org/10.1057/s41284-018-0154-1

Mazerolle, L. G., Frank, J., Rogan, D., & Watkins, C. (2000). *Field evaluation of the ShotSpotter gunshot location system: Final report on the Redwood City field trial*. U.S. Department of Justice.

Mazerolle, L. G., Watkins, C., Rogan, D., & Frank, J. (1998). Using gunshot detection systems in police departments: The impact on police response times and officer workloads. *Police Quarterly, 1*(2), 21–49. https://doi.org/10.1177/109861119800100202

McCollister, K. E., French, M. T., & Fang, H. (2010). The cost of crime to society: New crime- specific estimates for policy and program evaluation. *Drug Alcohol Dependency, 108*(1-2), 98–109. https://doi.org/10.1016/j.drugalcdep.2009.12.002

McEwen, T., & Regoeczi, W. (2015). Forensic evidence in homicide investigations and prosecutions. *Journal of Forensic Sciences, 60*(5), 1188–1198. https://doi.org/10.1111/1556-4029.12787

National Institute of Justice (NIJ). (2007). Spot the shot. *National Law Enforcement and Corrections Technology Center, TechBeat,* winter edition.

New York City Comptroller (NYCC). (2024). *Audit report on the New York City Police Department's oversight of its agreement with ShotSpotter Inc. for the gunshot detection and location system*. NYCC.

Onofrio, R. (2024). Amplifying school security with gunshot detection systems. *Security Industry Association*. https://www.securityindustry.org/2024/03/19/amplifying-school-security-with-gunshot-detection-systems/

O'Brien, D. T. (2024). "Big data" and evidence-based policy and practice: The advantages, challenges, and long-term potential of naturally occurring data. In B. C. Welsh, S. N. Zane, & D. P. Mears (Eds.), *The Oxford handbook of evidence-based crime and justice policy* (1st ed., pp. 598–620). Oxford University Press.

Papachristos, A. V., & Bastomski, S. (2018). Connected in crime: The enduring effect of neighborhood networks on the spatial patterning of violence. *American Journal of Sociology, 124*(2), 517–568. https://doi.org/10.1086/699217

Papachristos, A., Braga, A., Piza, E., & Grossman, L. (2015). The company you keep? The spillover effects of gang membership on individual gunshot victimization in social networks. *Criminology, 53*(4), 624–649. https://doi.org/10.1111/1745 9125.12091

Parish, W. J., Keyes, V., Beadles, C., & Kandilov, A. (2018). Using entropy balancing to strengthen an observational cohort study design: Lessons learned from an evaluation of a complex multi-state federal demonstration. *Health Services and Outcomes Research Methodology, 18*(1), 17–46. https://doi.org/10.1007/s10742-017-0174-z

Piza, E. L. (2012). *Using Poisson and negative binominal regression models to measure the influence of risk on crime incident counts*. Rutgers University, Rutgers Center on Public Security.

Piza, E., Arietti, R., Carter, J. & Mohler, G. (2025b). The effect of gunshot detection technology on evidence collection and case clearance in Kansas City, Missouri. *Journal of Experimental Criminology, 21*(1), 127–152.

Piza, E. L., Hatten, D. N., Carter, J. G., Baughman, J. H., & Mohler, G. O. (2023). Gunshot detection technology time savings and spatial precision: An exploratory analysis in Kansas City. *Policing: A Journal of Policy and Practice, 17*. https://doi.org/10.1093/police/paac097

Piza, E., Hatten, D., Mohler, G., Carter, J., & Cho, J. (2024a). Gunshot detection technology effect on gun violence in Kansas City, Missouri: A microsynthetic control evaluation. *Criminology & Public Policy, 23*(4), 863–892.

Piza, E., Mohler, G., Connealy, N., Arietti, R., & Carter, J. (2025a). Space-time association between gunshot detection alerts, calls for service, and police enforcement in Chicago: Differences across citizen race and incident type. *Journal of Quantitative Criminology, 41*(1), 53–74.

Piza, E. L., Mohler, G. O., Hatten, D. N., Connealy, N. T., Arietti, R., Cho, J., & Castillo, E. (2024b). *The impact of gunshot detection technology on gun violence in Kansas City and Chicago: A multi-pronged evaluation*. Office of Justice Programs, National Institute of Justice.

Piza, E. L., Welsh, B. C., Farrington, D. P., & Thomas, A. L. (2019). CCTV surveillance for crime prevention: A 40-year systematic review with meta-analysis. *Criminology & Public Policy, 18*(1), 135–159. https://doi.org/10.1111/1745-9133.12419

Piza, E. L., Wheeler, A. P., Connealy, N. T., & Feng, S. Q. (2020). Crime control effects of a police substation within a business-improvement district: A quasi-experimental synthetic control evaluation. *Criminology & Public Policy, 19*(2), 653–684. https://doi.org/10.1111/1745 9133.12488

Policing Project. (2019). *Privacy audit and assessment of ShotSpotter, Inc.'s gunshot detection technology.* New York University School of Law. https://www.policingproject.org/shotspotter

Ratcliffe, J. H. (2015). Towards an index for harm-focused policing. *Policing, 9*(2), 164–182. https://doi.org/10.1093/police/pau032

Ratcliffe, J. H., Lattanzio, M., Kikuchi, G., & Thomas, K. (2019). A partially randomized field experiment on the effect of an acoustic gunshot detection system on police incident reports. *Journal of Experimental Criminology, 15,* 67–76. https://doi.org/10.1007/s11292-018-9339-1

Reid, W. (1975). *Microphone array location from sounding made by a passing projectile.* Naval Surface Weapons Center.

Renda, W., & Zhang, C. H. (2019). Comparative analysis of firearm discharge recorded by gunshot detection technology and calls for service in Louisville, Kentucky. *International Journal of Geo-Information, 8,* 275–291. https://doi.org/10.3390/ijgi8060275

Robbins, M. W., Saunders, J., & Kilmer, B. (2017). A framework for synthetic control methods with high-dimensional, micro-level data: Evaluating a neighborhood-specific crime intervention. *Journal of the American Statistical Association, 112*(517), 109–126. https://doi.org/10.1080/01621459.2016.1213634

Sandau, A. G., Field, R., & Hunter, K. (2024). Detect, dispatch, drive: A study of ShotSpotter acoustic technology and transport of gunshot victims. *Journal of Surgical Research, 300,* 550–558. https://doi.org/10.1016/j.jss.2024.04.076

Saunders, J., Lundberg, R., Braga, A. A., Ridgeway, G., & Miles, J. (2015). A synthetic control approach to evaluating place-based crime interventions. *Journal of Quantitative Criminology, 31*(3), 413–434. https://doi.org/10.1007/s10940-014-9226-5

Shadish, S. R., Cook, T. D., & Campbell, D. T. (2002). *Experimental and quasi-experimental designs for generalized causal inference.* Houghton Mifflin.

Strom, K. J., Hendrix, J., Taniguchi, T., Aagaard, B., Werth, R. S., & Legacy, S. (2016). *Research on the impact of technology on policing strategy in the 21st century: Final report.* Office of Justice Programs.

Tonry, M. (2015). The fog around cost-of-crime studies may finally be clearing: Prisoners and their kids suffer too. *Criminology & Public Policy,* 14(4), 653–671. https://doi.org/10.1111/1745-9133.12170

Topper, M., & Ferrazares, T. (2023). *The effect of ShotSpotter technology on police response times.* [Working Paper, University of California, Santa Barbara]. University of California.

Vandergriff, C. (2016, February 12). CMPD cancels contract for ShotSpotter gunshot detection system. *Spectrum News 1.* https://spectrumlocalnews.com/tx/austin/news/2016/02/12/cmpd-cancels-contract-for-shotspotter-gunshot-detection-system

Verbeek, M. (2008). *A guide to modern econometrics* (3rd ed.). John Wiley.

Vovak, H., Riddle, T., Taniguchi, T., Hoogesteyn, K., & Yang, Y. (2021). *Strategies for Policing Innovation (SPI) in Wilmington, Delaware: Targeting violent crime* [Final Evaluation Report Prepared for the Wilmington Police Department and the Bureau of Justice Assistance].

Wandling, M. W., Avery, N. B., Shapiro, M. B., & Haut, E. (2016). Police transport versus ground EMS: A trauma system-level evaluation of prehospital care policies and their effect on clinical outcomes. *Journal of Trauma Care and Acute Care Surgery, 81*(5), 931–935. https://doi.org/10.1097/TA.0000000000001228

Watkins, C., Mazerolle, L. G., Rogan, D., & Frank, J. (2002). Technological approaches to controlling random gunfire: Results of a gunshot detection system field test. *Policing: An International Journal of Police Strategies & Management, 25*(2), 345–370. https://doi.org/10.1108/13639510210429400

Wheeler, A. P. (2020). Allocating police resources while limiting racial inequality. *Justice Quarterly, 37*(5), 842–868.

Wheeler, A. P., Gerell, M., & Yoo, Y. (2020). Testing the spatial accuracy of address-based geocoding for gunshot locations. *The Professional Geographer, 72*(3), 398–410. https://doi.org/10.1080/00330124.2020.1730195

WLKY. (2018, March 16). ShotSpotter selects Louisville as test city thanks to drone gunshot detection program. *WLKY.* https://www.wlky.com/article/shotspotter-selects-louisville-as-test-city-thanks-to-drone-gunshot-detection-program/19455500

Wooldridge, J. M. (2010). *Econometric analysis of cross section and panel data* (2nd ed.). MIT Press.

Zhao, Q., & Percival, D. (2017). Entropy balancing is doubly robust. *Journal of Causal Inference, 5*(1), 20160010. https://doi.org/10.1515/jci-2016-0010

4

Measuring and Studying Officer-Involved Shootings in the US: Comparing Estimates from Multiple Open-Source Registries

Justin Nix, Ian Adams, and Joshua McCrain

Historically, a lack of comprehensive data on police shootings has hindered a nuanced understanding of this critical issue. However, in the past decade, the research landscape has rapidly evolved with the emergence of improved data facilitated by crowd-sourced and non-governmental initiatives. This chapter explores the transformative potential of such data, reviewing recent findings that have significantly advanced our collective knowledge on the topic. We demonstrate numerous sources of bias in previous empirical work focusing on the effects of police shootings. Further, we emphasize the caution required when navigating this newfound wealth of information, outlining some of the inherent challenges and potential pitfalls associated with the analysis of crowdsourced data. Our aim is to contribute to more informed public and academic dialogues on police shootings.

4.1 Introduction

Roughly 1,000 people are killed by police gunfire each year in the United States – or about three each day, on average. Males comprise 95% of those killed. The most common decedent is white, but in per capita terms, Black people are disproportionately killed (Edwards et al., 2019; Mentch, 2020). Most decedents were armed with a firearm, cutting instrument, or blunt object (Nix et al., 2017), and about one in four displayed signs of mental illness in the moments leading up to their death (Saleh et al., 2018). Unsurprisingly, there is substantial geographic variation as well, with western and midwestern metropolitan statistical areas demonstrating higher rates of fatal police shootings than their northeastern counterparts (Schwartz & Jahn, 2020).

 DOI: 10.1201/9781003430773-5

Just a decade ago, we could not have opened our chapter this way, because at the time, no one knew precisely how often police shot and killed people, let alone situational patterns and trends therein. Although it was understood that the FBI's official data on "justifiable homicides" was incomplete (Fyfe, 2002; Klinger, 2012; Zimring, 2017), unofficial data compiled by Fatal Encounters, Mapping Police Violence, and *The Washington Post* emerged in the mid-2010s and showed just how inadequate the FBI's data were. These more comprehensive data have certainly been a welcome addition, but they are not without their own limitations.

In this chapter, we provide a brief overview of each dataset (plus a few others) and discuss challenges and pitfalls associated with unofficial data being the "go-to" source for information about police shootings. We wish to be clear at the outset that we have set aside discussion of deaths attributed to conducted energy weapons (e.g., TASERs; see White et al., 2013), vehicular pursuits (Alpert & Lum, 2014), restraint asphyxiation (Karch, 2016), and other causes – and instead focus strictly on police shootings, which significantly outnumber other such police deadly force incidents (Schwartz & Jahn, 2020; Nix, 2020). We also exclude from our discussion various state-level databases, whether official (e.g., California's Use of Force Incident Reporting System) or unofficial (e.g., *Tampa Bay Times'* data on police shootings in Florida from 2009 to 2014).

4.2 Unofficial Police Shooting Data: A Review of the Landscape

We are aware of seven "unofficial" databases (i.e., not maintained by the US government) that provide estimates of the number of police shootings in the United States: *Fatal Force*, *Mapping Police Violence*, *Fatal Encounters*, *Gun Violence Archive*, *The Counted*, *SPOTLITE*, and *PoliceShootingsData.com*.[1] Table 4.1 provides more information about each dataset, including what it sets out to track as well as the number of police shootings it estimates from 2015 to 2020.

There are several takeaways from Table 4.1. For starters, each database has slightly different inclusion criteria, which have implications for both the type and number of police shootings that they observe. On the more restrictive end of the continuum, *The Washington Post*'s "Fatal Force" database applies the most conservative definition, strictly tracking fatal shootings by on-duty police officers. The benefit of such a narrow definition is that it leaves little room for disagreement about the officer's intent to use deadly force. Of course, the drawback is that it excludes the highly publicized and consequential deaths of Eric Garner, Freddie Gray, and George Floyd, to

TABLE 4.1

Seven Unofficial Databases That Track and Publicly Share Information About Police Shootings

Dataset	Years Included	Creator(s)	Inclusion Criteria	# of Fatal Police Shootings, 2015–2020[a]	# of Nonfatal Police Shootings, 2015–2020[a]
Fatal Force	2015–present	*The Washington Post*	Incidents in which a police officer, in the line of duty, shoots and kills a civilian	5,946	Not included
Mapping Police Violence	2013–present	Mapping Police Violence, Inc.	All people who die "as a result of being shot, beaten, restrained, intentionally hit by a police vehicle, pepper sprayed, tasered, or otherwise harmed by police officers, whether on-duty or off-duty"	6,187	Not included
Fatal Encounters	2000–2022	D. Brian Burghart	All people "killed during interactions with police"	6,505	Not included
Gun Violence Archive	2014–present	Michael Klein and Mark Bryant	"all types of shootings are included, whether OIS, accidental, children shooting themselves, murders, armed robberies, familicide, mass shootings, DGU, Home Invasions, drivebys and everything else"	5,790[b]	4,595[b, c]
The Counted	2015–2016	*The Guardian*	"Any deaths arising directly from encounters with law enforcement. This will inevitably include, but will likely not be limited to, people who were shot, tasered and struck by police vehicles as well those who died in police custody."	NA	Not included
SPOTLITE	2014–2021	Cline Center for Advanced Social Research	"any incident where police use firearms—including those with nonfatal outcomes—as well as any other use of force that results in a death"	6,201	Included, but unclear in current version

(*Continued*)

TABLE 4.1 (CONTINUED)

Seven Unofficial Databases That Track and Publicly Share Information About Police Shootings

Dataset	Years Included	Creator(s)	Inclusion Criteria	# of Fatal Police Shootings, 2015–2020[a]	# of Nonfatal Police Shootings, 2015–2020[a]
PoliceShootingsData .com	1970–2024d	Clark et al. (2025)	Fatal and nonfatal police shootings in the 302 largest US cities (i.e., has 100,000+ residents and their own police department)	1,091	1,437[d]

[a]*Unless otherwise noted, these calculations were performed by the authors. For Fatal Force, Mapping Police Violence, and Gun Violence Archive, we restricted to fatal shootings by on-duty officers (as Fatal Force does not track off-duty shootings, and we were interested in making as close to apples-to-apples comparisons as possible). Fatal Encounters and SPOTLITE do not include a filter for on/off-duty status of the officer(s) involved, so we expect these totals inevitably include a small number of off-duty shootings. Any errors are ours.*

[b]*According to Ward et al. (2024, Table C2, "on-duty" + "on- and off-duty" totals).*

[c]*Note that this number refers to persons nonfatally injured by police gunfire. It does not include people whom officers shot at but did not injure.*

[d]*Note that the authors only have OIS data for a handful of agencies dating back to 1970. Note also that for this row, the "nonfatal" column does include people whom police shot at and missed.*

name but a few. None of these individuals were shot by police officers, and so their cases would not be tracked by *The Washington Post*, yet few would argue that they were not killed as a result of police action. At the other end of the continuum is the Gun Violence Archive, which attempts to track *all* instances of gun violence in the United States, including the small subset that involves police officers, whether as perpetrators or victims (Sierra-Arévalo et al., 2023).[2] Between these two extremes is Fatal Encounters, which tracks all deaths that occur during police-citizen interactions. Thus, it *does* capture the deaths of Garner, Gray, Floyd, and others who were not killed by gunfire. But it also includes, for example, thousands of individuals who committed suicide during standoffs with police officers.[3]

It is also noteworthy that the databases were launched by people from a variety of different backgrounds. Two of the databases – Fatal Force and The Counted – were created by teams of journalists (i.e., *The Washington Post* and *The Guardian*). Fatal Encounters was created by an individual, D. Brian Burghart, who had previously been a newspaper editor and journalism instructor. Meanwhile, SPOTLITE was created by a team of researchers at the University of Illinois' Cline Center for Advanced Social Research (Jennings et al., 2023), and PoliceShootingsData.com was created by a team of political scientists for their book *Deadly Force: Police Shootings in Urban America* (Clark et al., 2025). Gun Violence Archive was launched by Mark Bryant, a "shooting enthusiast" who "fell into the job by accident" (Cheves, 2017), and finally, members of the team that operates Mapping Police Violence also co-founded Campaign Zero, an activist group whose stated goal is to "live in a world beyond policing." We raise this point not to advocate for one database over others but to caution that each was created with its own goals or agendas in mind. The diverse origins of each source underscore the importance of critically evaluating each database's methodology, potential biases, and limitations. While all these efforts contribute valuable data to the field, users should approach the information with an awareness of the varying perspectives and objectives that may have shaped data collection and presentation.

Table 4.1's headline counts might appear tightly clustered to some readers. Only 715 deaths separate the lowest (Gun Violence Archive, 5,790) and highest (Mapping Police Violence, 6,505) estimates for 2015–2020 (roughly, an average of 120 per year). Yet even this modest spread can redirect statistical conclusions. Because fatal police shootings are statistically rare and scholars often analyze thin slices of the data (for example, unarmed victims, race-specific victims, or single city-year strata), adding or subtracting a few incidents reshapes denominators, alters subgroup composition, and can meaningfully shift coefficients.

We emphasize this issue because both over-inclusion and omission of cases demonstrably change findings. Nix and Lozada (2021) re-coded the cases used in Bor et al. (2018) and showed that 91 of the cases arguably lay outside the study's own definition of a police killing of an unarmed subject. Deleting those events reduced every reported coefficient and rendered the headline claim that "police killings of unarmed Black Americans have effects on mental health among Black American adults in the general population" (p. 302) statistically null. A similar audit of Legewie (2019) uncovered numerous classification errors; once corrected, the purported spill-over from police killings to adverse Black infant health vanished, prompting voluntary retraction. Definitions also drive contradictory inferences: appearing just one month apart in top political-science journals, Markarian (2023) and Morris and Shoub (2023) each reported a significant causal link between police killings and voter turnout, but in opposite directions, a divergence traceable to different inclusion rules and database snapshots. These examples arose in widely read medical, general-science, and top field journals that attract post-publication scrutiny; the incidence of similar problems in less-examined venues is unknowable, though the underlying dependence on shifting public databases is identical.

In the last column of Table 4.1, we attempted to identify the number of shootings by on-duty police officers logged by each database from 2015 to 2020. Alas, this was an imperfect exercise as we could not easily filter shootings by off-duty officers in the Fatal Encounters and SPOTLITE datasets. Nevertheless, even among those where we could restrict to on-duty shootings – Fatal Force, Mapping Police Violence, and Gun Violence Archive – we observed notable discrepancies (see also Comer & Ingram, 2023). Gun Violence Archive logged 5,790 fatal shootings, while Fatal Force and Mapping Police Violence logged 5,946 and 6,187, respectively. Ultimately, these discrepancies might boil down to minor differences in other inclusion criteria (e.g., whether to include shootings by federal agents), duplicates, and/or a small number of false positives and negatives. Regardless, the differences highlighted in Table 4.1 should concern researchers attempting to study Officer-Involved Shootings (OIS) – a statistically rare outcome – or subsets of OIS that are even rarer (e.g., police shootings of women, unarmed individuals, juveniles, or persons displaying signs of mental illness).

One final takeaway from Table 4.1 is that only three of the seven databases – Gun Violence Archive, SPOTLITE, and PoliceShootingsData.com – attempt to track nonfatal police shootings. Gun Violence Archive suggests that over 40% of people struck by police gunfire from 2015 to 2020 did not die (n = 4,595). PoliceShootingsData.com estimates that during the same period in 208 mid to large US cities, upwards of 56% of people shot at by police officers did not die (n = 1,437).[4] Indeed, the omission of nonfatal police shootings from the other databases amounts to a significant limitation. As the late criminologist James Fyfe, a pioneer in the study of OIS, observed:

> *Deadly force is physical force capable of or likely to kill; it does not always kill. The true frequency of police decisions to employ firearms as a means of deadly*

force, therefore, can best be determined by considering woundings and off-target shots as only fortuitous variations of fatal shootings. All are of a kind.

(1978, p. 32, emphasis in original)

When using their firearms, police officers are trained to aim at their target's torso (as it is the largest and therefore easiest target to hit) and to continue firing until the threat has been neutralized (Clark et al., 2025). Thus, to strictly track people killed by police officers (by any means) is to track a nonrandom subset of less than half the universe of deadly force incidents (Baćak et al., 2021; Klinger, 2012; Klinger et al., 2016; McCann et al., 2017; Ward et al., 2024; White, 2006).[5] In other words, "deadly force" is only sometimes deadly, and the characteristics of people killed by police may not adequately resemble the characteristics of people who were the recipients of police deadly force (see, e.g., Clark et al., 2025; Nix & Shjarback, 2021; Ward et al., 2024). This, in turn, frustrates efforts by researchers to measure the causes and impacts of various environmental, organizational, officer, and civilian characteristics on police use of deadly force (Nix, 2020).

In fact, the majority of what we think we know about police use of deadly force comes from published studies that did not include any information on nonfatal police shootings. But it turns out this is just one of many potential threats to the validity of such studies – a point we expand upon in the next section.

4.3 Draw the DAG

A Directed Acyclic Graph (DAG) is an essential tool for mapping the complex causal relationships that influence the occurrence and outcomes of officer-involved shootings (OIS). By visually representing these relationships, a DAG clarifies the pathways through which various factors interact to produce specific outcomes. The strength of this approach lies in its ability to identify confounding variables, highlight potential causal mechanisms, indicate sources of bias, and guide the selection of appropriately specified statistical models (Pearl & Mackenzie, 2018). DAGs allow researchers to precisely state their estimand and relate it to their statistical model.

While DAGs are important for any inferential research, we argue they are especially important for studying police use of force incidence. This particular use case, and the illustration we create below, demonstrates that both the data-generating process *and* the causal question are downstream of important, non-trivial assumptions. Without the structured approach DAGs facilitate, researchers may be generating quantities of interest related to OIS that are biased (with unknown magnitude and direction) – or, in the worst case, quantities entirely unrelated to their actual empirical objective.

In the context of policing and the use of deadly force, a DAG enables researchers to systematically explore how different levels of influence – ranging from organizational policies to community characteristics – intersect to affect police-civilian encounters and their outcomes and what we are able to observe in the data. In other words, *the dataset itself is downstream of a complex causal process* (see Knox, Lowe, and Mummolo, 2020). This approach enables researchers to determine which features of the policing environment are downstream to their independent variable(s) ("post-treatment") and thus would introduce bias when included in a regression. To facilitate this analysis, we have identified the key nodes that should be included in the DAG, representing the various factors that contribute to and result from OIS incidents:

1. **Organizational Context**: Factors such as staffing levels, training, culture, policies, and deployment patterns that influence officer characteristics and behaviors, as well as the likelihood of an OIS directly.
2. **Officer Characteristics and Behaviors**: The attributes and actions of officers that directly impact the nature of police-civilian encounters and the likelihood of an OIS occurring, as well as the OIS's likelihood of being covered by news media or mentioned on social media.
3. **Civilian Characteristics and Behaviors**: The attributes and actions of civilians that directly affect police-civilian encounters and the likelihood of an OIS occurring, as well as the OIS's likelihood of being covered by news media or mentioned on social media.
4. **Community Context**: Environmental factors such as 911 call volume, crime rates, poverty, segregation, and (dis)trust in police that influence civilian characteristics and behaviors, as well as the likelihood of police-civilian encounters, and news or social media coverage should an OIS occur. Community context can also indirectly influence the outcome of an OIS through its direct relationship with the healthcare environment.
5. **Police-Civilian Encounter**: The interaction between officers and civilians, influenced by all the above factors, which can lead to an OIS.
6. **Outcome of OIS**: The possible results of an OIS, which can include fatality, nonfatal injury, or no injury.
7. **Healthcare Environment**: The community's healthcare infrastructure, such as the availability and quality of trauma centers, can influence the outcome of an OIS.
8. **Media Coverage**: The extent to which an OIS is mentioned in news media or on social media is influenced by the OIS outcome and other

factors, including characteristics and behaviors of both the officer(s) and civilian(s), and both the organizational and community contexts.

9. **Database Inclusion**: The likelihood of an OIS being captured in unofficial databases like those maintained by *The Washington Post*, Mapping Police Violence, Fatal Encounters, Gun Violence Archive, SPOTLITE, and others, often driven primarily by media coverage.

By mapping these nodes, we gain a clearer understanding of the pathways leading to OIS and their subsequent documentation.

Each item in the list above becomes a "node" in the DAG, and the researcher's task then is to map out the relationships among them. Figure 4.1 is our attempt to map out the factors known to influence the likelihood of an OIS, and in turn, the factors that influence an OIS being captured by any of the aforementioned unofficial databases. To analyze these pathways effectively, we conceptually divide the DAG into two primary phases: "Before the Trigger Is Pulled" and "After the Trigger Is Pulled." This division allows for a more nuanced examination of the distinct causal mechanisms at play in each phase.

4.3.1 Before the Trigger Is Pulled

Before the trigger is pulled, a confluence of factors establishes the conditions under which an OIS may occur. Organizational context plays a foundational role, influencing both officer behavior and departmental strategies. Key elements such as staffing levels, training protocols, organizational culture, and deployment patterns directly affect how officers engage with civilians (Ingram et al., 2018; Mourtgos et al., 2024; Sierra-Arévalo, 2024). For example, agencies that implement comprehensive use-of-force policies and prioritize de-escalation training tend to experience fewer incidents of excessive force (Engel et al., 2022) and OIS (Jennings & Rubado, 2017; Shjarback et al., 2021).

The community context also exerts a significant influence on the likelihood of an OIS. Variables such as crime rates, levels of poverty, 911 call volume, and the degree of community trust in the police shape both officer and civilian behavior and, by extension, the nature of police-civilian encounters (Bottoms & Tankebe, 2012; Fyfe, 1980; Klinger et al., 2016; MacDonald et al., 2001; Maksuta et al., 2024; Nix et al., 2020; Smith & Alpert, 2007). In areas with high levels of crime, concentrated disadvantage, and/or strained police-community relations, these encounters are often more frequent and volatile, increasing the potential for escalation to an OIS (Keller et al., 2023).

During police-civilian interactions, the behaviors and decisions of both parties are influenced by a host of factors. Civilian demeanor, perceived threat, and compliance play critical roles in shaping how officers respond (Alpert et al., 2020; Nix et al., 2019). At the same time, officers' stress levels and their ability to make decisions under pressure can determine whether a situation escalates into violence (James et al., 2019). The presence of multiple officers further complicates these dynamics, with evidence suggesting that "contagious fire" can occur when one officer's use of force triggers others to fire as well, independent of the situation's initial threat level (DeCarlo et al., 2024).

4.3.2 After the Trigger Is Pulled

Once the trigger is pulled, the focus shifts to the immediate and long-term consequences of the OIS. The outcome – whether it results in a fatality, nonfatal injury, or no injury at all – is influenced by various factors beyond the shooting itself. One crucial determinant is the surrounding healthcare environment. The availability and quality of trauma care in the vicinity can significantly impact whether an injury from an OIS leads to death (Crandall et al., 2013; Hou, 2022). This aspect is often overlooked in existing literature, which predominantly focuses on fatalities, potentially missing more than 50% of all instances of lethal force by officers (Baćak et al., 2021; Klinger, 2012; Klinger et al., 2016; McCann et al., 2017; Ward et al., 2024; White, 2006).

The aftermath of an OIS also involves a complex web of causal factors that determine how the incident is perceived and recorded. Media coverage is a key element in this phase, influencing public awareness and the inclusion of incidents in various databases. Factors such as the race of the civilian involved (Schroedel & Chin, 2020; Zuckerman et al., 2019), the circumstances of the shooting (Moreno-Medina et al., 2022), and the broader sociopolitical context can affect whether and how an OIS is reported in the news or on social media. This media coverage, in turn, largely determines whether an incident is captured in various unofficial databases.

This reliance on media reports introduces significant biases into the data collected by these databases. Incidents that result in fatalities or receive significant media attention are more likely to be documented, while nonfatal incidents and misses, or OIS occurring in areas with less local news coverage, are more likely to be overlooked (Baćak et al., 2021; Gobaud et al., 2023; Kaufman et al., 2020). This selective reporting skews the data, leading to an incomplete understanding of the true scope and nature of OIS incidents. For example, if one would like to study differences in OIS incident effects in rural versus urban areas, one must be willing to make the unrealistically strong assumptions that (a) the dataset is as complete in rural areas as in urban areas; and (b) if this is not the case, the missingness is as-if random. Given the differences in news coverage, policies, transparency, reporting requirements, etc., each type of area would face, such an assumption is obviously insurmountable. By breaking down the DAG into the phases before and after the trigger is pulled, researchers can better account for these biases and develop more comprehensive strategies for studying and addressing police use of deadly force.

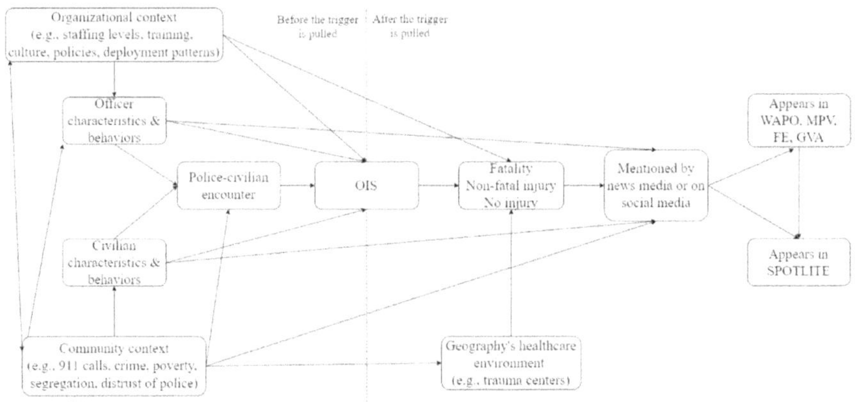

FIGURE 4.1
Factors That Influence the Likelihood of a Police Shooting and Its Inclusion in Various Unofficial Datasets.

4.3.3 Cross-Over Relationships

Our before-and-after structure is not intended to suggest that there are no causal considerations that cross over that period. Indeed, an additional challenge in studying this topic is that some nodes cross over the "trigger line," meaning they influence both the likelihood of an OIS occurring and the way that OIS is subsequently recorded and reported. These cross-over nodes represent complex, multi-directional relationships that likely require more interdisciplinary research to fully understand. For instance, community characteristics such as crime rates, socioeconomic status, and the level of trust in law enforcement not only affect the dynamics of police-civilian encounters (Nix et al., 2020) but also influence the amount and nature of media coverage these incidents receive. Similarly, organizational characteristics – such as departmental transparency and fairness (Trinkner et al., 2016), public relations strategies (Cheng, 2023), and the presence of body-worn cameras (Lum et al., 2020; White & Malm, 2020) – can shape both officer behavior during an encounter and how the incident is portrayed in the media (Cheng, 2021; Tregle et al., 2022).

These cross-over relationships are particularly challenging to study because they straddle the line between pre- and post-incident factors, making it difficult to disentangle their effects. To better understand these dynamics, scholars need to draw on expertise from fields such as media studies, sociology, and communications, in addition to criminology and policing research. By fostering interdisciplinary collaborations, researchers can develop more nuanced theories and models that account for these complex interactions, ultimately leading to a more comprehensive understanding of the factors that influence both the occurrence of OIS and their representation in public discourse.

4.4 Empirical Implications

One could argue that the above DAG is incomplete or inconsistent with a particular worldview. One could similarly state that the world is much more complex than the DAG presented here, and in fact, there are many other observed or unobserved factors unspecified in this stylized model. Further, one could tell a straightforward story about how police shootings in time *t-1* predict police shootings in time *t*. These are all valid points and again emphasize the importance of the research being explicit about the assumptions made in their implied (or explicitly stated) causal model. Adjusting the DAG to account for these very reasonable critiques would only serve to make it more complex.

How do these considerations matter for your regression specification? Specifically, there are numerous concerns that the applied research should be acutely aware of. First, making the DAG even more complex than the one presented here would introduce more confounding (much of which is unobserved and thus not in our data), and indicate more sources of post-treatment bias. In the case of confounding, unless the regression specification includes all confounders, it will always be biased (with potentially unknown size if the confounders are unobserved).

For a concrete example, imagine your research question includes the relationship between how officers interact with the public and the propensity of police use of force. An obvious confounder in this relationship is public mood/opinion towards policing in a given community, because that shared sentiment shapes both sides of the encounter: residents who distrust the police are more likely to display non-compliance, verbal resistance, or hostile body language (Engel, 2003), while officers who anticipate such hostility shorten their verbal warning sequence, escalate commands, and are quicker to resort to force (Nix & Pickett, 2017; Nix, Pickett, & Mitchell, 2019; Van Maanen, 1978). We may be able to capture a coarse survey-based measure of quantity and include it in the regression, potentially alleviating some concern over this source of bias. However, that quantity may not exist or even be measurable at all; in such a case, this regression will always be biased. Similarly, we may be interested in how department policies and procedures predict police use of force. An obvious quantity downstream from department policy is how/whether officers interact with individuals in a community. If we include some measure capturing that officer-civilian interaction in a regression, we are introducing post-treatment bias into our regression.

4.5 Concluding Thoughts

This chapter has examined the complexities of OIS and the challenges in accurately capturing and analyzing related data. Using a Directed Acyclic Graph (DAG) approach, we emphasized the importance of considering both the factors leading up to an OIS ("Before the Trigger Is Pulled") and those influencing its aftermath ("After the Trigger Is Pulled").

One key takeaway is the significant limitation in relying solely on fatal OIS data. Ward et al. (2024) demonstrate that fatalities constitute only about 55% of all injurious police shootings (i.e., excluding when officers shoot and miss). This undercount skews our understanding of police use of lethal force and overlooks crucial factors that determine whether an OIS results in a fatality, injury, or no harm (Nix & Shjarback, 2021).

While unofficial databases have filled a critical gap in understanding police shootings, they are not without limitations. Scholars must approach these data with caution, ensuring their analyses are grounded in a comprehensive understanding of the factors influencing both the occurrence of OIS and its documentation. Only through such rigorous research can we advance knowledge and contribute meaningfully to discussions on police use of force.

We have summarized how the factors leading up to an OIS involve multiple critical variables, including organizational context, officer and civilian characteristics, and community context. Equally important are the factors that dictate whether an OIS, once it occurs, is captured in various crowdsourced databases. The reliance on non-random samples driven by media coverage and other post-incident factors introduces biases that can distort scholarly analysis and public understanding. To address these critical shortcomings in the literature, we advocate for the explicit use of DAGs in research on police shootings. A well-constructed DAG provides a clear framework for identifying causal relationships and confounding variables and forces the researcher to confront potential issues in their data-generating process. "Showing your DAG" enhances transparency and rigor in analysis, ensuring that assumptions are clearly stated and evaluated.

Much of the research produced since 2014 has been limited by dependence on these non-random samples. It is essential for researchers to critically assess the databases they use and be clear about any filtering or cleaning of the data performed prior to analysis. Scholars must be wary of studies that engage superficially with these datasets without accounting for their inherent complexities and biases. For scholars looking to impact an important area of research, this may come as good news – dozens, if not hundreds, of study designs are available and ready to be tested under better conditions. The Gun Violence Archive and SPOTLITE datasets, for example, provide researchers with information about nonfatal police shootings, thereby providing a more complete account of deadly force incidents in the United States.

By leveraging more rigorous methods, expanding statistical models to include nonfatal shootings, including the use of DAGs, and a critical approach to data sources, researchers can address the significant gaps left by earlier studies. This presents an opportunity to produce more accurate, comprehensive insights into police use of force, ultimately contributing to more informed policy decisions and better outcomes in policing practices.

Notes

1. SPOTLITE is an acronym for "Systematic Policing Oversight Through Lethal-force Incident Tracking Environment."
2. Unfortunately, it is not easy to filter out all the gun violence that does not involve police officers. Incidents can be tagged with a variety of characteristics (e.g., "officer-involved shooting," "mass shooting"), requiring users to manually review incidents that have multiple tags. As but one example, there are 500 rows in the database pertaining to the 2017 Las Vegas mass shooting, and they all include a tag for "officer-involved shooting."
3. Another example: in a since-deleted tweet, @FatalEncounters announced on May 26, 2022, that each of the 19 victims of the Uvalde school shooting would be added to the database, since officers were on scene. Whether such incidents amount to "fatal encounters" between police and civilians is clearly more subjective than incidents involving the fatal shooting of a civilian by a police officer.
4. Note that the outcome was undetermined for ~20% of shootings over this period (n = 617), so the true number of nonfatal shootings is undoubtedly higher.
5. Clark et al. (2025) show systematic differences in what predicts a fatality conditional on being shot.

References

Alpert, GeoJrey P., and Cynthia Lum. *Police Pursuit Driving: Policy and Research.* Springer Science & Business Media, 2014.Alpert, GeoJrey, Roger Dunham, Justin Nix, Kyle McLean, and Scott Wolfe. "Authority maintenance theory of police-citizen interactions." In *Critical Issues in Policing: Contemporary Readings.* edited by Roger G. Dunham, Geoffrey P. Alpert, and Kyle D. McLean, 8th ed. Long Grove: Waveland Press, 2020: 376–397.

Baćak, Valerio, Joshua Gary Mausolf, and Christiane Schwarz. "How comprehensive are media-based data on police officer–involved shootings?" *Journal of Interpersonal Violence* 36, nos. 17–18 (2021): NP10055-NP10065.

Bor, J., A. S. Venkataramani, D. R. Williams, and A. C. Tsai. "Police killings and their spillover effects on the mental health of black Americans: A population-based, quasi-experimental study." *The Lancet* 392, no. 10144 (2018): 302–310.

Bottoms, Anthony, and Justice Tankebe. "Beyond procedural justice: A dialogic approach to legitimacy in criminal justice." *Journal of Criminal Law and Criminology* 102 (2012): 119.

Cheng, Tony. "Social media, socialization, and pursuing legitimation of police violence." *Criminology* 59, no. 3 (2021): 391–418.

Cheng, Tony. *The Policing Machine: Enforcement, Endorsements, and the Illusion of Public Input*. University of Chicago Press, 2023.

Cheves, John. "Website counts the bodies as they fall to gunfire across America." *Lexington Herald-Leader* (2017). Retrieved from <https://www.thestar.com/news/world/website-counts-the-bodies-as-they-fall-to-gunfire-across-america/article_b60804d4-d624-58a9-a4a1-a7c59ee1d5bc.html>.

Clark, Tom, Adam Glynn, and Michael Leo Owens. *Deadly Force: Police Shootings in Urban America*. Princeton University Press, 2025.

Comer, Benjamin P., and Jason R. Ingram. "Comparing fatal encounters, mapping police violence, and Washington Post fatal police shooting data from 2015–2019: A research note." *Criminal Justice Review* 48, no. 2 (2023): 249–261.

Crandall, Marie, Douglas Sharp, Erin Unger, David Straus, Karen Brasel, Renee Hsia, and Thomas Esposito. "Trauma deserts: Distance from a trauma center, transport times, and mortality from gunshot wounds in Chicago." *American Journal of Public Health* 103, no. 6 (2013): 1103–1109.

DeCarlo, John, Eric Dlugolenski, and David Myers. "An experimental test of the contagious fire thesis in policing." *Journal of Criminal Justice* 93 (2024): 102215.

Edwards, Frank, Hedwig Lee, and Michael Esposito. "Risk of being killed by police use of force in the United States by age, race–ethnicity, and sex." *Proceedings of the National Academy of Sciences* 116, no. 34 (2019): 16793–16798.

Engel, Robin S. Explaining suspects' resistance and disrespect toward police. *Journal of Criminal Justice, 31*(5), (2003): 475–492.

Engel, Robin S., Nicholas Corsaro, Gabrielle T. Isaza, and Hannah D. McManus. "Assessing the impact of de-escalation training on police behavior: Reducing police use of force in the Louisville, KY Metro Police Department." *Criminology & Public Policy* 21, no. 2 (2022): 199–233.

Fyfe, James J. *Shots Fired: An Examination of New York City Police Firearms Discharges*. PhD diss., State University of New York at Albany, 1978.

Fyfe, James J. "Geographic correlates of police shooting: A microanalysis." *Journal of Research in Crime and Delinquency* 17, no. 1 (1980): 101–113.

Fyfe, James J. "Too many missing cases: Holes in our knowledge about police use of force." *Justice Research and Policy* 4, no. 1–2 (2002): 87–102.

Gobaud, Ariana N., Christina A. Mehranbod, Elinore Kaufman, Jonathan Jay, Jessica H. Beard, Sara F. Jacoby, Charles C. Branas, Brady Bushover, and Christopher N. Morrison. "Assessing the gun violence archive as an epidemiologic data source for community firearm violence in 4 US cities." *JAMA Network Open* 6, no. 6 (2023): e2316545–e2316545.

Hou, Yuchen. *Fatal and Non-Fatal Police Shootings in the United States, 2015: An Examination of Open-Source Data*. PhD diss., City University of New York, 2022.

Ingram, Jason R., William Terrill, and Eugene A. Paoline III. "Police culture and officer behavior: Application of a multilevel framework." *Criminology* 56, no. 4 (2018): 780–811.

James, Lois, Stephen James, Rachel Davis, and Elizabeth Dotson. "Using interval-level metrics to investigate situational-, suspect-, and officer-level predictors of police performance during encounters with the public." *Police Quarterly* 22, no. 4 (2019): 452–480.

Jennings, Jay T., and Meghan E. Rubado. "Preventing the use of deadly force: The relationship between police agency policies and rates of officer-involved gun deaths." *Public Administration Review* 77, no. 2 (2017): 217–226.

Jennings, Jay, Ajay Singh, Scott Althaus, Michael Martin, Joseph Bajjalieh, and Jennifer Robbennolt. "Cline center SPOTLITE United States dataset." *University of Illinois at Urbana-Champaign*, 2023. Retrieved from <https://clinecenter.illinois.edu/spotlite/data>.

Karch, Steven B. "The problem of police-related cardiac arrest." *Journal of Forensic and Legal Medicine* 41 (2016): 36–41.

Kaufman, Elinore J., Jesse E. Passman, Sara F. Jacoby, Daniel N. Holena, Mark J. Seamon, Jim MacMillan, and Jessica H. Beard. "Making the news: Victim characteristics associated with media reporting on firearm injury." *Preventive Medicine* 141 (2020): 106275.

Keller, Tyler, Joel Caplan, and Leslie Kennedy. "The influence of the environment on officer-involved shootings." *Homicide Studies* 27, no. 1 (2023): 97–119.

Klinger, David A. "On the problems and promise of research on lethal police violence: A research note." *Homicide Studies* 16, no. 1 (2012): 78–96.

Klinger, David, Richard Rosenfeld, Daniel Isom, and Michael Deckard. "Race, crime, and the micro-ecology of deadly force." *Criminology & Public Policy* 15, no. 1 (2016): 193–222.

Knox, Dean, Will Lowe, and Jonathan Mummolo. "Administrative records mask racially biased policing." *American Political Science Review* 114, no. 3 (2020): 619–637.

Legewie, J. "Police violence and the health of black infants." *Science Advances* 5, no. 12 (2019): eaax7894.

Lum, Cynthia, Christopher S. Koper, David B. Wilson, Megan Stoltz, Michael Goodier, Elizabeth Eggins, Angela Higginson, and Lorraine Mazerolle. "Body-worn cameras' effects on police officers and citizen behavior: A systematic review." *Campbell Systematic Reviews* 16, no. 3 (2020).

MacDonald, John M., Robert J. Kaminski, GeoJrey P. Alpert, and Abraham N. Tennenbaum. "The temporal relationship between police killings of civilians and criminal homicide: A refined version of the danger-perception theory." *Crime & Delinquency* 47, no. 2 (2001): 155–172.

Maksuta, Kyle D., Yunhan Zhao, and Tse-Chuan Yang. "Race, disadvantage, and violence: A spatial exploration of macrolevel covariates of police-involved homicides within and between US counties." *Social Science Research* 119 (2024): 102985.

Markarian, G. A. "The impact of police killings on proximal voter turnout." *American Politics Research* 51, no. 3 (2023): 414–430.

McCann, A., K. Hamilton, T. Dolven, CaulderwoodK. , and R. Arthur. "Police shoot far more people than anyone realized, a VICE News investigation reveals." *Vice [Internet]* 12 (2017).

Mentch, Lucas. "On racial disparities in recent fatal police shootings." *Statistics and Public Policy* 7, no. 1 (2020): 9–18.

Moreno-Medina, Jonathan, Aurelie Ouss, Patrick Bayer, and Bocar A. Ba. "Officer-involved: The media language of police killings." No. w30209. National Bureau of Economic Research, 2022.

Morris, K. T., and K. Shoub. "Contested killings: The mobilizing effects of community contact with police violence." *American Political Science Review* (2023): 1–17.

Mourtgos, Scott M., Ian T. Adams, and Justin Nix. "Staffing levels are the most important factor influencing police response times." *Policing: A Journal of Policy and Practice* 18 (2024): paae002.

Nix, Justin. "On the challenges associated with the study of police use of deadly force in the United States: A response to Schwartz & Jahn." *PLOS ONE* 15, no. 7 (2020): e0236158.

Nix, Justin, Bradley A. Campbell, Edward H. Byers, and GeoJrey P. Alpert. "A bird's eye view of civilians killed by police in 2015: Further evidence of implicit bias." *Criminology & Public Policy* 16, no. 1 (2017): 309–340.

Nix, J., and Pickett, J. T. Third-person perceptions, hostile media effects, and policing: Developing a theoretical framework for assessing the Ferguson effect. *Journal of Criminal Justice, 51,* (2017): 24–33.

Nix, J., and M. J. Lozada. "Police killings of unarmed black Americans: A reassessment of community mental health spillover effects." *Police Practice and Research* 22, no. 3 (2021): 1330–1339.

Nix, Justin, Justin T. Pickett, and Renée J. Mitchell. "Compliance, noncompliance, and the in-between: Causal effects of civilian demeanor on police officers' cognitions and emotions." *Journal of Experimental Criminology* 15 (2019): 611–639.

Nix, Justin, Justin T. Pickett, and Scott E. Wolfe. "Testing a theoretical model of perceived audience legitimacy: The neglected linkage in the dialogic model of police–community relations." *Journal of Research in Crime and Delinquency* 57, no. 2 (2020): 217–259.

Nix, Justin, and John A. Shjarback. "Factors associated with police shooting mortality: A focus on race and a plea for more comprehensive data." *PLOS ONE* 16, no. 11 (2021): e0259024.

Pearl, Judea, and Dana Mackenzie. *The Book of Why: The New Science of Cause and ePect.* Basic books, 2018.

Saleh, Amam Z., Paul S. Appelbaum, Xiaoyu Liu, T. Scott Stroup, and Melanie Wall. "Deaths of people with mental illness during interactions with law enforcement." *International Journal of Law and Psychiatry* 58 (2018): 110–116.

Schwartz, Gabriel L., and Jaquelyn L. Jahn. "Mapping fatal police violence across US metropolitan areas: Overall rates and racial/ethnic inequities, 2013–2017." *PLOS ONE* 15, no. 6 (2020): e0229686.

Schroedel, Jean Reith, and Roger J. Chin. "Whose lives matter: The media's failure to cover police use of lethal force against Native Americans." *Race and Justice* 10, no. 2 (2020): 150–175.

Shjarback, John A., Michael D. White, and Stephen A. Bishopp. "Can police shootings be reduced by requiring officers to document when they point firearms at citizens?" *Injury Prevention* 27, no. 6 (2021): 508–513.

Sierra-Arévalo, Michael. *The Danger Imperative: Violence, Death, and the Soul of Policing.* Columbia University Press, 2024.

Sierra-Arévalo, Michael, Justin Nix, and Scott M. Mourtgos. "The 'war on cops,' retaliatory violence, and the murder of George Floyd." *Criminology* 61, no. 3 (2023): 389–420.

Smith, Michael R., and GeoJrey P. Alpert. "Explaining police bias: A theory of social conditioning and illusory correlation." *Criminal Justice and Behavior* 34, no. 10 (2007): 1262–1283.

Tregle, Brandon, Justin Nix, and Justin T. Pickett. "Body-worn cameras and transparency: Experimental evidence of inconsistency in police executive decision-making." *Justice Quarterly* 39, no. 3 (2022): 455–477.

Trinkner, Rick, Tom R. Tyler, and Phillip Atiba GoJ. "Justice from within: The relations between a procedurally just organizational climate and police organizational efficiency, endorsement of democratic policing, and officer well-being." *Psychology, Public Policy, and Law* 22, no. 2 (2016): 158.

Van Maanen, John. ""The asshole."" In *Policing: A View from the Street*, edited by Peter K. Manning and John Van Maanen. Pacific Palisades, CA: Goodyear Publishing Company, (1978): 221–238.

Ward, Julie A., Javier Cepeda, Dylan B. Jackson, Odis Johnson Jr, Daniel W. Webster, and Cassandra K. Crifasi. "National burden of injury and deaths from shootings by police in the United States, 2015–2020." *American Journal of Public Health* 114, no. 4 (2024): 387–397.

White, Michael D. "Hitting the target (or not): Comparing characteristics of fatal, injurious, and noninjurious police shootings." *Police Quarterly* 9, no. 3 (2006): 303–330.

White, Michael D., Justin Ready, Courtney Riggs, Donald M. Dawes, Andrew Hinz, and JeJrey D. Ho. "An incident-level profile of TASER device deployments in arrest-related deaths." *Police Quarterly* 16, no. 1 (2013): 85–112.

White, Michael D., and Aili Malm. *Cops, Cameras, and Crisis: The Potential and the Perils of Police Body-worn Cameras*. NYU Press, 2020.

Zimring, Franklin E. *When Police Kill*. Harvard University Press, 2017.

Zuckerman, Ethan, J. Nathan Matias, Rahul Bhargava, Fernando Bermejo, and Allan Ko. "Whose death matters? A quantitative analysis of media attention to deaths of Black Americans in police confrontations, 2013–2016." *International Journal of Communication (Online)* (2019): 4751–4778.

5

On Racial Bias and Fatal Police Shootings: Insights from Publicly Reported Data

Lucas Mentch and Ryan Cecil

5.1 Introduction

The extensive media coverage of fatal police shootings in the United States in recent years has ignited polarizing debates, sparked widespread controversy, and triggered global political movements. Most notably, there have been a number of high-profile incidents in recent years in which Black Americans seem to have died at the hands of police unnecessarily. This increased attention on policing and use of force has also fueled claims of a racial disparity in police use of force, especially lethal force. Raw population-level data would seem to back this up; between January 2015 and April 2024, nearly 27% of police shooting victims[1] were Black despite making up little more than 13% of the overall population [22]. In that same timespan, whites made up more than 75% of the nationwide population but less than 51% of the total number of police shooting victims [[illegible], 31]; see Figure 5.1 for a full accounting. If factors such as crime density, the racial distribution of victims and perpetrators of crime, overall racial demographics, and the density of employed law enforcement officers (LEOs) could be assumed to be roughly uniform across the US, then these high-level summary statistics alone might be enough to suggest a potential underlying racial bias in the use of fatal force by police. At the very least, this information may lead one to reasonably infer that the racial composition and overall numbers of fatal police shootings deviate from what would be anticipated if an officer's decision to use lethal force were independent of race. However, such assumptions are inherently unreasonable for a vast and diverse nation like the United States.

Besides being highly emotional and oftentimes graphic, the topic of police shootings remains politically contentious in the US for a number of reasons. First, official data on police use of force has historically been incredibly

 DOI: 10.1201/9781003430773-6

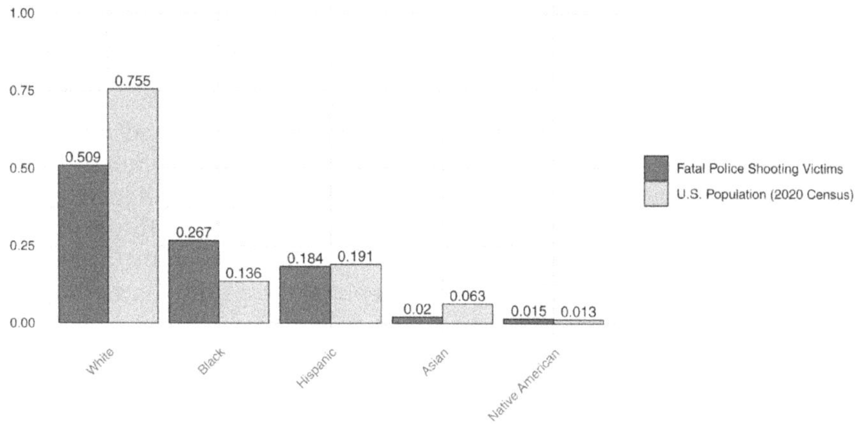

FIGURE 5.1
Racial Proportions by Total US Population vs. Fatal Shooting Victims. Fatal shooting proportions were tabulated directly from the data provided by *The Washington Post* [31] dataset; population proportions are shown according to the US Census Bureau [29].

murky and difficult, if not impossible, to access. Despite urgings from both Congress and the President, federal reporting, even on lethal use of force, has traditionally been voluntary for individual police departments to provide. Indeed, the National Use of Force Data Collection Program was only initiated by the FBI in 2015 and was opened to begin receiving reports in 2019 [8, 12]. Reporting rates have been unsurprisingly dismal with only 27% of police departments nationwide submitting reports; in states such as New York and Pennsylvania, only a small handful of departments have submitted data, covering only about 2% of officers in their respective states [12]. Both underreporting and potentially biased data reporting on this topic have been widely documented in the academic literature as well [6, 11, 13, 14, 15, 17, 23, 32, 33].

As a result, numerous organizations, including the *Washington Post* [31] and the *Guardian* [28], have taken it upon themselves to independently create databases of fatal police shooting incidents via the use of web crawlers designed to find, extract, and aggregate relevant information on police shootings from local media reports. In 2015, then head of the FBI James Comey called it *"embarrassing and ridiculous"* that the federal government has no better information on police shootings than databases compiled by the *Guardian* US and the *Washington Post* [25]. This same sentiment was echoed soon after in an academic research article [15], where it was claimed that "the best data on police use of force come to us not from the government or from scholars, but from the Washington Post."

But while the lack of a respected and accessible centralized data repository has hindered both the quality and quantity of research on the topic, researchers have nonetheless still managed to produce substantial amount of literature; for a small sample of notable research from the past two decades, see for example, [1–4, 9, 10, 18 – 20, 24, 26, 27], or, for a more comprehensive overview, see recent literature reviews by Ridgeway (2016) [26] and Nix et al. (2017) [23]. In more recent years, researchers have begun focusing on the more complete data provided by the large journalistic outlets noted above. For example, Nix et al. (2017) [23] utilized the *Washington Post* data to examine "threat-perception failures" and found racial disparities in the likelihood that an individual was attacking an officer and/or armed at the time of the shooting. Klinger and Slocum (2017) [15] disputed these findings, noting that even unarmed individuals can pose a potentially serious threat.

Disputes such as these help illustrate an additional, perhaps even more significant reason for the continued controversy surrounding police shootings: even within the academic literature – and certainly to a much greater extent in popular media – we find seemingly conflicting, near contradictory reports on the matter. Consider two articles, both published in June 2020. One, an op-ed published in the *Wall Street Journal* was titled "The Myth of Systemic Police Racism" and claimed that there is no evidence of widespread racial bias on the part of the police, citing a number of scholarly articles [21]. The other, published at fivethirtyeight.com, also cited numerous studies and yet claimed that *"The data seems to overwhelmingly point to a criminal justice system riven by racial bias. But, remarkably, it could be even more overwhelming than some studies make it seem"* [5]. Both articles are written in an authoritative fashion, giving the impression that these conclusions are definitive and backed by scientific research, yet the conclusions themselves seem almost diametrically opposed. One could forgive a member of the general public on either side of the political isle for feeling as though their "side" is correct and backed by settled science. We'll return to elaborate more on this issue in Section 5.3.

The primary purpose of this chapter is to expand upon earlier work by Mentch (2020) [22] in looking at the racial distributions of fatal police shooting victims. Put simply, we follow [22] in asking: *Taking into account local characteristics, how many individuals from each race would be reasonable to expect if the fatal shooting victims could be seen as a random sample from the localized population and are the observed victim totals in line with such expectations?* As discussed in the introductory paragraph, because the US constitutes such a diverse collection of areas in terms of things like crime and arrest rates, police employment rates, and population density and diversity, it is crucial to take these local characteristics into account when trying to accurately assess the degree to which the observed racial distribution of fatal police shooting victims is unexpected. To address this primary question of interest, we utilize recent data on fatal police shootings obtained through the *Washington Post* [31], as well as datasets containing county-level racial demographics from the US Census Bureau [29], along with data on law enforcement density [7], and local arrest demographics [30]. Note that the *Washington Post* database is utilized in its entirety here – we make no attempt to distinguish between (or weight any differently) shootings classified as "justified" or "non-justified," nor shootings where the suspect(s) are armed/unarmed, etc. For the purposes of the analyses in the following sections, each fatal police shooting incident on which we have data is treated in the same fashion, as simply one location where such an incident occurred.[2]

The remainder of this chapter is laid out as follows. In Section 5.2 we present the main analytical results, demonstrating that the racial distribution of fatal police shooting victims is far from what would be expected relative to local population demographics, but much more in line with expectations relative to local arrestee demographics. We return to the question of why there is so much apparent disagreement in the literature on this topic in Section 5.3 before concluding with a summary in Section 5.4.

5.2 Racial Demographics of Shooting Victims

Recall our primary analytical question of interest: given the number of fatal police shooting victims in recent years, how many victims are reasonable to expect from each race and are the observed totals in line with those expectations? It is readily apparent from high-level summary statistics that the racial makeup of these victims differs drastically from overall nationwide racial population demographics. However, police employment tends to be higher in areas with larger populations, which are oftentimes more racially diverse. Consequently, assuming a positive correlation between the number of officers and the incidence of shootings, one might expect the pool of fatal police shooting victims to be more racially diverse than indicated by nationwide population demographics. We now employ a series of randomized experiments and simulations to investigate this idea under various settings and assumptions; readers seeking a more thorough and technical discussion of these kinds of experiments are invited to see [22].

Here and in the experiments below, we utilize the *Washington Post* data [31] and note that a total of 9,593 fatal police shooting victims are recorded between January 2015 and April 2024. Among these cases, a total of 8,329 contain sufficient information to be included in the initial analyses below.[3] Our first analysis is arguably the most straightforward: for each of the 8,460 fatal police shootings with sufficient information recorded by the *Washington Post,* we simply cross reference this with racial demographics of that local area provided by the 2020 US Census [29] and sample a race at random weighted accordingly. In other words, following the procedure in [22], if a shooting occurred in a county where the population is 50% White, 30% Black, 5%

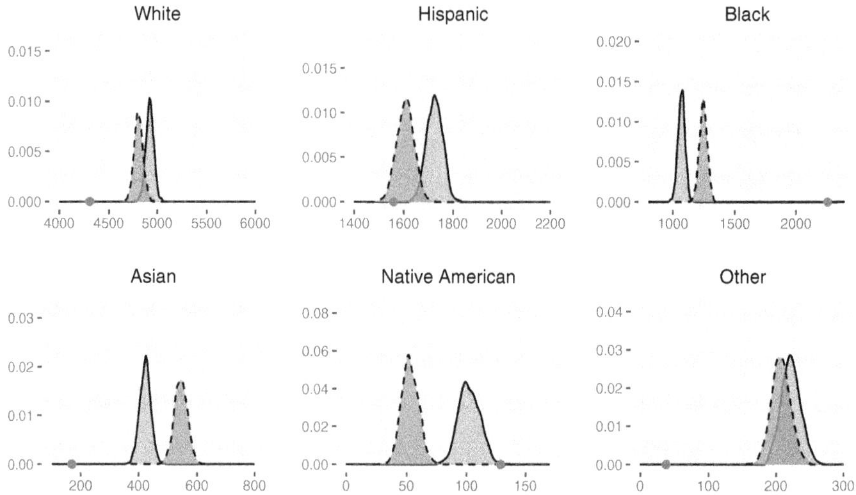

FIGURE 5.2
Density Estimates for Victim Totals by Race Assuming Fixed Shooting Locations (Solid Line) or Random Locations (Dashed Line) Relative to the 2020 Census Data on Local Racial Demographics. Points along each horizontal axis correspond to the observed victim totals.

Native American, 5% Asian, and 10% Other, then the probabilities of selecting a White, Black, Native American, Asian, or Other individual as a hypothetical victim would be 0.5, 0.3, 0.05, 0.05, and 0.10, respectively. We then repeat this for each shooting location, sum the number of victims of each race to get an estimated victim total for each race, and then repeat the entire process 1000 times to get a distribution of plausible victim totals by race and investigate where the observed victim total falls with respect to that distribution. Those density estimates by race are shown as a solid line in Figure 5.2 with the observed victim totals shown as a gray dot along the x-axis.

5.2.1 Shooting Locations as Random

This analysis, though simple, is also quite naive. We are, in essence, simulating a distribution of fatal police shooting victims under the assumption that such victims can be seen as simply chosen at random from the local population where the shootings occurred. To enhance things a bit, consider the fact that there is likely a fair amount of randomness in the shooting locations themselves. As observed in [22], for example, if police pursue a suspect across multiple counties and the incident ends in a fatal shooting, that incident is recorded as having happened in the county where the pursuit concluded, disregarding information about the county where it began and any other counties involved. Furthermore, during the same timeframe, many other unrelated *nonfatal* shootings may have occurred under similar circumstances, but their nonfatal nature precludes them from being included in the *Washington Post* dataset.

Suppose then that we consider the shooting locations as random. That is, rather than sampling directly from only those local populations where fatal shootings were recorded, we instead see those locations as a random sample from all locations where fatal shootings may have been most likely to occur. To do this, we first generate the location at random weighted according to local police employment [7] so that counties employing more police are (proportionally) more likely to be selected. Note that this is equivalent to simply selecting an officer uniformly at random from the population of all US police officers. Once a location is randomly generated, a victim race is then randomly generated according to the local population demographics in exactly the same fashion as before. This process is then repeated 8,460 times – once for each fatal shooting recorded in the *Washington Post* dataset – and the number of hypothetical victims from each race is then totaled. The entire procedure is then repeated 2000 times so that for each race we have 2000 associated hypothetical victim counts. The resulting distributions of those victim counts for each race under this random location setup are shown as a dashed line in Figure 5.2.

Looking at Figure 5.2, first note that for most races, the switch from fixed to random shooting locations – though significant for some races – is not overly dramatic. For White, Hispanics, and Native Americans, the distributions arising under the random location setup (dashed line) are shifted a bit left, suggesting that those races are "over-sampled" relative to the locations where fatal police shootings actually occurred. For Asians and Blacks, the distributions shift right, and for the "Other" racial category, the distributions remain largely the same. More pertinently, for every race except Hispanic – and possibly Native Americans to a lesser degree – it is plain to see that the observed victim totals (shown in gray points on the x-axis) differ wildly from what would be expected relative to either simulated distribution (i.e., regardless of the kind of simulation employed). In fact, if we were to formally quantify this following the empirical p-value calculation procedure laid out in [22], all p-values would be 0 for the White, Black, Asian, and Other races. In other words, for Hispanics and Native Americans, the number of victims appears at least somewhat in line with what might be expected, but there is very strong evidence that the victim totals for Whites, Blacks, Asians, and Others would be exceedingly unlikely to have been observed by random chance if the simulated distributions were accurate; the observed victim counts for Whites, Asians, and Others are significantly less than what would be expected while the number of Black victims dramatically exceeds expectations.

We pause here to highlight some high-level inferences that can be drawn from the preceding two analyses. From Figure 5.2, we see immediately that for most races, the observed victim totals (points shown in gray on the x-axis) lie quite far from the distributions of expected victim totals (shown under the solid line curve). This tells us that it is exceedingly unlikely that police shooting victims constitute a simple random sample from the local populations where those shootings occurred. In the second analysis, we treated those shooting locations as random and considered the possibility that fatal police shootings were more likely to occur in areas where more police officers were employed. Instead of sampling victims at random from local populations where shootings were observed, we instead sampled victims proportionally more often from local populations where more officers were employed. Here too, however, we see much the same story – the observed victim totals (gray points on the x-axis) lie quite far from these distributions of expected victim totals (shown under the dashed line) under a random location setup. This tells us that even focusing our sampling more intensely on areas of highest police employment is thoroughly insufficient for explaining the wide disparities in observed versus expected victim totals by race.

5.2.2 Sampling from the Local Arrestee Population

These analyses reveal that factoring in local racial demographics of where police shootings either have taken place or are most likely to take place offers little to no help in explaining the vast disparities in the racial distribution of fatal police shooting victims. Those simulation setups, however, remain quite naive: they operate as if all individuals in the local population are equally likely to interact with and ultimately be shot and killed by police.

Suppose instead that we assumed that (1) individuals who interact most with police are also most likely to become fatal police shooting victims and (2) the rate at which police interact with members of various races is proportional or at least similar to the rate at which police *arrest* members of those races. Under this alternative setup, we could repeat the above simulations and sample hypothetical victim races relative to the local *arrestee* demographics [30] rather than the local population demographics. In particular, these experiments can be performed under the same fixed and random location designs described above. In the fixed location setting, victims are sampled relative to the local arrestee population where shootings actually occurred; under the random location setup, victims are sampled relative to the local arrestee population where shootings may be most likely to occur based on police employment figures. The results under both fixed and random location setups[4] are shown in Figure 5.3.

Looking at Figure 5.3, we notice immediately that the observed victim totals now appear much more in line with what would be expected when sampling from local arrestee populations rather than the local civilian populations that informed the distributions in Figure 5.2. Table 5.1 provides coarse summaries of the outcomes from each of the experiments conducted in this work, which helps to further highlight this fact. In the table, note that results pertaining to the fixed location setups are shown on the left and those found under the random location setups on the right. Four columns appear under each setup: Demo, Arrest, Shift, and Fit Imp.

The Demo columns show the number of standard deviations the observed victim counts for each race fell from the expected count (under the respective

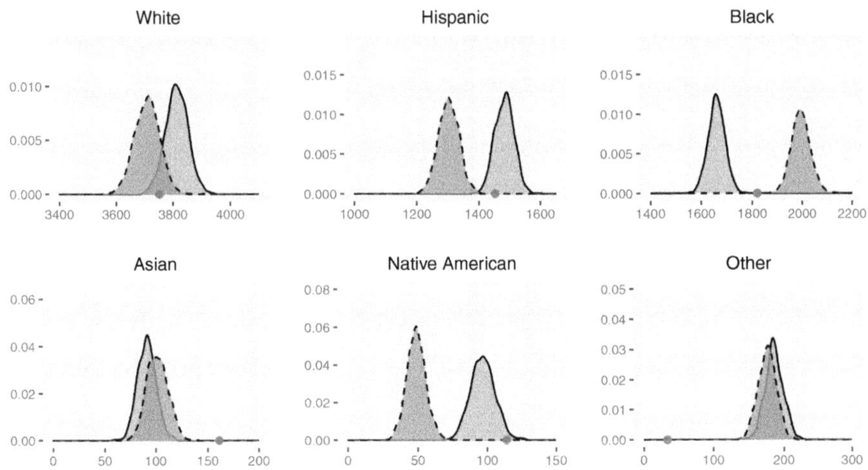

FIGURE 5.3
Density Estimates for Victim Totals by Race Assuming Fixed Shooting Locations (Solid Lines) or Random Locations (Dashed Lines) Relative to the Local Arrestee Demographics. Points along each horizontal axis correspond to the observed victim totals.

fixed and random location setups) whenever victims were sampled according to local population demographics. Similarly, the Arrest columns show the number of standard deviations the observed victim counts for each race fell from the expected count whenever victims were sampled according to local arrestee populations. The Shift columns indicate the number of standard deviations by which the distributions shift when sampling from the arrestee populations rather than the civilian populations. The fit improvement (Fit Imp) columns show the degree to which the fit of those distributions is improved (i.e., how many standard deviations closer the observed victim count for each race lies to the expected count when sampling from the local arrestee population rather than the local population at large). More formally, letting Demo_r and Arrest_r denote the standard deviation differences for race r, under both the fixed and random location sampling schemes, the shift and fit improvements are defined as

$$\begin{aligned} \text{Shift}_r &= |\,\text{Demo}_r - \text{Arrest}_r| \\ \text{Fit Imp} &= |\,\text{Demo}_r| - |\,\text{Arrest}_r|. \end{aligned}$$

A few outcomes in Table 5.1 are worth highlighting. Under the fixed location setup, note that the observed victim total for Whites lies nearly 15 standard deviations below what would be expected relative to the demographic distribution, while that of Blacks lies more than 40 standard deviations above the expected. This fact alone is worth pausing to digest. These are enormously large standard deviations; the probability of observing this by random chance is essentially 0. Equally remarkable, however, is the fact that these

TABLE 5.1

Number of Standard Deviations Away that the Observed Victims Totals Lie from the Center of the Empirical Distributions under the Fixed and Random Location Setups with Respect to Local Demographics (**Demo** Columns) and Local Arrestee Populations (**Arrest** Columns)

	Fixed Location				Random Location			
	Demo	**Arrest**	**Shift**	**Fit Imp**	**Demo**	**Arrest**	**Shift**	**Fit Imp**
White	–15.10	–1.56	13.54	+13.54	–10.98	1.01	11.99	+9.97
Hisp	–5.06	–0.98	4.08	+4.08	–1.34	4.59	5.93	–3.25
Black	41.85	4.83	37.02	+37.02	31.51	–4.58	36.09	+26.93
Asian	–13.50	7.63	21.13	+5.87	–16.88	5.79	22.67	+11.09
NatAm	2.99	1.97	1.02	+1.02	10.76	9.55	1.21	+1.21
Other	–13.06	–11.57	1.49	+1.49	–12.02	–11.09	0.93	+0.93

Note: Also included is the number of standard deviations the distributions shift when sampling from the arrestee populations vs from the overall local demographics (**Shift** Columns) as well as the relative improvement in fit (**Fit Imp** Columns).

distributions shift so dramatically – by more than 13 and 37 standard deviations for Whites and Blacks, respectively – when sampling from arrestee rather than civilian populations. The same story holds for the random location setup, albeit to a less dramatic degree, as well as for most other races under both the fixed and random location settings. The only small exception is the random location setup for Hispanics, in which the observed victim total lies a few standard deviations further when sampling from arrestee demographics, though we note here that relative to the others the fit for Hispanics is already quite good. For Native Americans and Others, the fit improves under arrestee sampling, but to a relatively small degree.

Finally, we close this section by stressing that the closer agreement between observed and expected victim counts seen here by sampling from the arrestee population does not itself imply or even necessarily suggest a lack of racial bias on the part of police. Rather, it suggests only that any racial biases present on the level of arrests largely carries through to the level of fatal shootings.

5.3 Shortcomings and Controversies

The analyses above, culminating in Table 5.1, tell a fairly straightforward story. For some races, e.g., Hispanics and Native Americans, the number of fatal police shooting victims is largely in line with what would be expected. For others, e.g., Whites and Blacks especially, there are far fewer (or more, respectively) victims than would be expected relative to the local civilian populations where fatal shootings either occurred or are likely to occur. However, for such races, the observed victim count becomes substantially more plausible when sampling relative to arrestee rates rather than civilian populations. Blacks constitute a particularly extreme example: demographic sampling under the fixed location setup suggests the observed victim total is more than 41 standard deviations higher than expected, yet arrestee sampling suggests the observed total is off by less than 5 standard deviations – a shift of more than 37 standard deviations. Virtually any statistical test with any reasonable amount of power whatsoever would easily confirm that the observed count is significantly different from expected when sampling from the civilian population, yet finding evidence of disparity when sampling from the arrestee population would be somewhat more challenging.[5]

These stark and dramatic differences in conclusions shed light on the striking variations in reporting on issues related to racial bias and police shootings. If one wants to argue in favor of a racial bias by police, one can simply appeal to the dramatic disparities between demographics and rates of fatal police shootings. When one wants to argue the opposite, one can make a similarly strong appeal to arrest rates. To see this explicitly, consider the

June 2020 oped referenced in the introduction in which Heather MacDonald states that the

> *share of black victims is less than what the black crime rate would predict, since police shootings are a function of how often officers encounter armed and violent suspects. In 2018, the latest year for which such data have been published, African-Americans made up 53% of known homicide offenders in the U.S. and commit about 60% of robberies, though they are 13% of the population.*
>
> **[21]**

Indeed, consider generic statements such as "Black men are more likely to be shot and killed by police" or "police shootings are racially biased." Despite these kinds of statements being vigorously debated throughout the media, there are myriad ways in which these general ideas can be scientifically formalized and statistically analyzed. When one makes the first claim, for example, do they mean to suggest only that Black men are disproportionately likely to be victims of fatal police shootings? That is, comparing across the entire populations of all Black men and all White men, for example, is there evidence that a significantly higher proportion of those Black men will die by fatal police shootings? The available data unequivocally support this. On the other hand, perhaps one means to imply that in many police interactions, the police would be more likely to fatally shoot and kill a Black suspect than they would be to fatally shoot and kill a White suspect who was acting in an identical fashion. Recent studies have cast serious doubt on this claim; see e.g. [10]. Nonetheless, regardless of the particular metrics under investigation, the results of studies on racial disparities and police use of force are very frequently bluntly summarized as either "there is" or "there is not" racial bias on the part of police. Even scientific articles and other academic writings can sometimes be unfortunately sloppy in this regard.

Adding to the confusion is the myriad of issues associated with the kinds of data needed to perform the above analyses. These analyses require four different sources of data: in addition to the information on fatal police shootings, we also need information on local racial demographics, police employment rates, and local arrest rates by race. Even merging these datasets is a nontrivial task. Some datasets contain information at the county level, while others contain only city information, some counties may be omitted entirely from certain datasets, the data may have been collected in different years, and each dataset may contain its own share of typos that need to be addressed. Different datasets may also treat the same variables in different ways. For example, the *Washington Post* dataset labels "Hispanic" as one possible race, while the others treat this as an ethnicity. We refer the interested reader to [22] for a full accounting of how those issues were dealt with here.

There are also a multitude of issues related to the police shooting data in particular. As alluded to above, this dataset contains information only on fatal shootings, whereas one might plausibly wish to include nonfatal shootings and other types of fatal police encounters in this kind of an analysis. There are also a number of variables recorded that necessarily involve some degree of subjectivity, such as the level of threat the victim was exhibiting prior to being shot and whether or not he or she was displaying signs of mental illness. Any studies utilizing these variables should be undertaken and interpreted with extreme care. Finally, we note that the *Washington Post* database is dynamic; information is added and updated periodically as it becomes available. While this would seem to be a positive if the information is continually being made more accurate, the downside is that it makes previous studies more difficult to replicate. It also highlights that even some variables that may seem objective – e.g. what object(s), if anything the victim was armed with – may be more subjective than assumed. Earlier versions of the *Washington Post* dataset, for example, listed particular suspects as being armed with things like cell phones and chairs whereas later versions list those same suspects as unarmed.

Finally, we close by highlighting a final issue that has received increased attention in recent years. As noted above, the vast majority of available data on police encounters contains information on only those very serious – typically fatal – incidents. Even when more thorough data is available at a local level, by definition, we can observe only whatever information is recorded. Recent work by Knox et al. [16] pointed out that, as a result, any racial bias present on the level of stops (e.g., if police choose to stop or interact with certain racial groups more often than others) is masked by such records.

5.4 Conclusion

The primary purpose of this chapter was to give readers a brief introduction to the kinds of public data available on fatal police shootings and demonstrate the kinds of insights that might be drawn from it. Our analyses looked at the total number of victims of fatal police shootings in recent years by race and attempted to compare those observed totals to what would be expected under various setups including the hypothetical sampling of victims at random from local civilian and arrestee populations based on where shootings actually occurred and also by sampling the shooting location itself at random weighted according to local police employment. By and large, our findings are in agreement with [22] in that the observed victim totals by race are more in line with what would be expected when sampling from the arrestee rather than the civilian populations. Furthermore, the races for which the observed victim counts are generally the most unexpected relative to local demographics – Blacks, Whites, and Asians – are also those where the most dramatic improvements are seen when hypothetical victims are sampled from the local arrestee populations where fatal shootings occurred.

As noted earlier, however, we again caution that the local arrestee populations are not necessarily the same as the local criminal populations and thus the findings here suggest that any biases present on the arrest level may carry through to the fatal encounter level. As discussed in the preceding section, the fact that expectations differ so drastically for some races depending on how hypothetical victims are sampled helps explain the seemingly contradictory conclusions often reached in both academic research and popular media articles. At the end of the day, whether or not the racial distribution of victims appears "reasonable" depends almost entirely on how that notion is defined. We urge any reader seeking an accurate understanding of the ongoing research in this area to examine the work in-depth in order to fully grasp how things are being measured and defined so that the conclusions and other high-level summaries of that work can be understood in the appropriate context.

Notes

1. We use the term "fatal police shooting *victim*" here only as shorthand to refer to one who has been shot and killed by police; the term *victim* is not intended to convey any underlying political leanings, nor imply that police necessarily acted wrongfully in applying lethal force in each situation.
2. Further discussion on issues related to the *Washington Post* data is provided in Section 5.3 of this chapter, as well as in Chapter 4 written by Justin Nix.
3. Fatal shooting instances were omitted from these analyses if the victim race was missing in the *Washington Post* data (1,133 instances). Once these instances were removed, an additional 131 shooting instances were then further removed as they occurred in counties where police employment data was missing. To our knowledge, other than a recency bias with respect to the missing race information in the *Washington Post* data, there are no systematic patterns in the missingness observed or other biases present in the final dataset utilized.
4. These simulations require matching the information across an additional dataset on arrestee demographics, and as a result, there are less observations with common information across all three datasets. After removing the initial 1,133 instances of missing race information in the *Washington Post* data, we then removed an additional 1,128 instances where arrestee demographics were missing for the county in which the shooting occurred. Lastly, we removed an additional 131 instances where police employment information was missing, bringing the total number of fatal shooting incidents to 7,332 for the purpose of these simulations.
5. In most contexts, a difference of more than 4 standard deviations would generally be considered quite strong evidence. However, it's important to keep in mind here that this is nearly 10 times smaller than when sampling from local demographics. Further, under the random location setup, these findings

suggest that the number of observed Black victims is more than 4 standard deviations *below* what would be expected when sampling from local arrestee populations.

References

1. Kenneth Adams, Geoffrey Alpert, Roger Dunham, Joel Garner, Lawrence A Greenfeld, Mark A Henriquez, Patrick A Langan, Christopher D Maxwell, and Steven K Smith. Use of force by police: Overview of national and local data. *US Washington, DC: Department of Justice, National Institute of Justice Research Report*, 1999.
2. Geoffrey P Alpert and Roger G Dunham. The force factor: Measuring police use of force relative to suspect resistance. Police Executive Research Forum Washington, DC, 1997.
3. Geoffrey P Alpert and Roger G Dunham. *Understanding Police Use of Force: Officers, Suspects, and Reciprocity*. Cambridge University Press, 2004.
4. Barak Ariel, William A Farrar, and Alex Sutherland. The effect of police body-worn cameras on use of force and citizens? Complaints against the police: A randomized controlled trial. *Journal of Quantitative Criminology*, 31(3):509–535, 2015.
5. Laura Bronner. Why statistics don't capture the full extent of the systemic bias in policing. *fivethirtyeight.com*.
6. Joseph Cesario, David J Johnson, and William Terrill. Is there evidence of racial disparity in police use of deadly force? Analyses of officer-involved fatal shootings in 2015–2016. *Social Psychological and Personality Science*, 10(5):586–595, 2019.
7. Federal Bureau of Investigation. FBI Uniform Crime Reporting Program. Table 80: Full-time Law Enforcement Employees, by State by Metropolitan and Nonmetropolitan Counties, 2011. https://ucr.fbi.gov/crime-in-the-u.s/2011/crime-in-the-u.s.-2011/tables/table-80/view. (Access Date: 07/12/2016).
8. Federal Bureau of Investigation. National Use-of-Force Data Collection. https://www.fbi.gov/how-we-can-help-you/more-fbi-services-and-information/ucr/use-of-force. (Access Date: 05/07/2024).
9. Lorie Fridell and Hyeyoung Lim. Assessing the racial aspects of police force using the implicit- and counter-bias perspectives. *Journal of Criminal Justice*, 44:36–48, 2016.
10. Roland G Fryer Jr. An empirical analysis of racial differences in police use of force. *Journal of Political Economy*, 127(3):1210–1261, 2019.
11. James J Fyfe. Too many missing cases: Holes in our knowledge about police use of force. *Justice Research and Policy*, 4(1–2):87–102, 2002.
12. Tom Jackman. For a second year, most U.S. police departments decline to share information on their use of force. *The Washington Post*.
13. David Klinger, Richard Rosenfeld, Daniel Isom, and Michael Deckard. Race, crime, and the micro-ecology of deadly force. *Criminology & Public Policy*, 15(1):193–222, 2016.

14. David A Klinger. On the problems and promise of research on lethal police violence: A research note. *Homicide Studies*, 16(1):78–96, 2012.
15. David A Klinger and Lee Ann Slocum. Critical assessment of an analysis of a journalistic compendium of citizens killed by police gunfire. *Criminology & Public Policy*, 16(1):349–362, 2017.
16. Dean Knox, Will Lowe, and Jonathan Mummolo. Administrative records mask racially biased policing. *American Political Science Review*, 114(3):619–637, 2020.
17. Arthur L Kobler. Figures (and perhaps some facts) on police killing of civilians in the united states, 1965–1969. *Journal of Social Issues*, 31(1):185–191, 1975.
18. Nicolien Kop and Martin C Euwema. Occupational stress and the use of force by dutch police officers. *Criminal Justice and Behavior*, 28(5):631–652, 2001.
19. Joscha Legewie and Jeffrey Fagan. Group threat, police officer diversity and the deadly use of police force. *Columbia Public Law Research Paper*, (14–512), 2016.
20. Hyeyoung Lim, Lorie A Fridell, and Hoon Lee. The impact of supervision and neighborhood context on police use of less-lethal force: A multi-level analysis. *Journal of Police Studies*, 14(2):155–182, 2014.
21. Heather Mac Donald. The myth of systemic police racism. *The Wall Street Journal*.
22. Lucas Mentch. On racial disparities in recent fatal police shootings. *Statistics and Public Policy*, 7(1):9–18, 2020.
23. Justin Nix, Bradley A Campbell, Edward H Byers, and Geoffrey P Alpert. A bird's eye view of civilians killed by police in 2015: Further evidence of implicit bias. *Criminology & Public Policy*, 16(1):309–340, 2017.
24. Eugene A Paoline III and William Terrill. Police education, experience, and the use of force. *Criminal Justice and Behavior*, 34(2):179–196, 2007.
25. Betsy Reed. Fbi chief: 'unacceptable' that guardian has better data on police violence. *The Guardian*.
26. Greg Ridgeway. Officer risk factors associated with police shootings: A matched case-control study. *Statistics and Public Policy*, 3(1):1–6, 2016.
27. Brad W Smith. Structural and organizational predictors of homicide by police. *Policing: An International Journal of Police Strategies & Management*, 27(4):539–557, 2004.
28. The Guardian. The Counted: People Killed by the Police in the US. https://www.theguardian.com/us-news/ng-interactive/2015/jun/01/the-counted-police-killings-us-database. (Access Date: 05/09/2024).
29. United States Census Bureau. Annual County Resident Population Estimates by Age, Sex, Race, and Hispanic Origin: April 1, 2020 to July 1, 2022. https://www2.census.gov/programs-surveys/popest/technical-documentation/file-layouts/2020-2022/cc-est2022-alldata.pdf. (Access Date: 05/07/2024).
30. United States Department of Justice. Federal Bureau of Investigation. Uniform Crime Reporting Program Data: Arrests by Age, Sex, Race, 2013. ICPSR36115v1. Ann Arbor, MI: Inter-university Consortium for Political, and Social Research [distributor], 2015-10-07. https://doi.org/10.3886/ICPSR36115.v1. (Access Date: 09/12/2017).
31. Washington Post. Washington Post Database of Police Shootings. https://www.washingtonpost.com/graphics/national/police-shootings/. (Access Date: 07/12/2016).

32. Michael D White. Transactional encounters, crisis-driven reform, and the potential for a national police deadly force database. *Criminology & Public Policy*, 15(1):223–235, 2016.
33. Howard E Williams, Scott W Bowman, and Jordan Taylor Jung. The limitations of government databases for analyzing fatal officer-involved shootings in the united states. *Criminal Justice Policy Review*, 30(2):201–222, 2019.

6

Learning About the Illegal Market for Ammunition through the Lens of Recovered Firearms and Ammunition

Charles E. Loeffler

6.1 Introduction

Gun violence represents the biggest public safety challenge facing the United States today (Gramlich 2023). In 2021, 21,000 individuals were murdered with firearms, and another 40,000 individuals were injured with firearms (GVA 2023). Some estimates put the toll of firearm victimization even higher (Kena and Truman 2022). A substantial portion of this safety challenge comes from the diversion of legally purchased firearms into the illegal firearm market, where they are then used to commit crimes ranging from armed robbery to homicide. This diversion, *despite* regulatory prohibitions and enforcement actions by the Bureau of Alcohol, Tobacco, and Firearms (ATF) and local law enforcement agencies, has focused scholarly attention on the pathways through which firearms make their way into the hands of prohibited possessors. To date, this scholarship has documented that the market for illegal firearms involves a mixture of theft, straw purchasing for prohibited end-users by non-prohibited intermediaries, brokerage through associates and other third parties, and, increasingly, the purchase of semi-finished firearms or firearm components from private manufacturers (Braga et al. 2002, 2021; Harlow et al. 2001; Pierce et al. 2004; Sheley and Wright 1993). In some jurisdictions, tracing of recovered firearms has revealed that many illegal firearms come from out-of-jurisdiction supply sources. In others, tracing has revealed that most recovered firearms come from within jurisdiction sources. This informal supply network, but especially the out-of-jurisdiction variant, has been alternatively characterized as an "iron pipeline" and a "diffuse network of capillaries," reflecting the diversity of supply routes observed

DOI: 10.1201/9781003430773-7

through tracing recovered firearms and interviews with market participants (Chesnut et al. 2017).

These alternative supply network conceptualizations, however, suggest a larger conceptual ambiguity: is the market for illegal firearms and ammunition best understood as a robust supply channel as reflected in the popular press, with such quotes as "Some criminals claim that it is as easy to buy a gun on the streets as it is to buy fast food" (Terry 1992), or is it a thin market, where buyers incur large transaction costs due to the difficulty of connecting interested buyers and willing sellers, as suggested by recent scholarship (Cook et al. 2007)?

A thin market, whether caused by information asymmetries, regulation, taxes, or lack of liquidity, makes it difficult for buyers and sellers to transact (Hayenga, Gardner, and Houck 1978; Lehmann 1938). Classic measures of market thinness include low volumes of transactions, high price volatility, and large gaps between what buyers wish to pay and what sellers expect to receive (Durlauf and Blume 2016). As a model for illegal ammunition markets, it offers a useful conceptual tool for making sense of high observed mark-ups between legal and illegal sale prices for the same good (Cook et al. 2007; Hales, Lewis, and Silverstone 2006). If the illegal market for ammunition is thin, then small regulatory or enforcement adjustments might exacerbate the already precarious situation for market participants, leading to a more dysfunctional illegal market but a safer world. In the extreme, these efforts might turn a thin market into a shortage market in which illegal firearm users, regardless of fiscal resources, struggle to obtain adequate ammunition. However, if the illegal ammunition market is sufficiently robust, then current regulatory and enforcement efforts may be inadequate, and regulators might need to rethink aspects of the current regulatory and enforcement paradigm.

Part of the challenge in adjudicating between these competing conceptualizations is that the primary empirical evidence on the difficulty that prohibited end-users have in obtaining ammunition, information which is used to both develop and validate market models, comes from specially designed and rarely administered interviews of special populations, including jail inmates, gun brokers, and gang members. These populations are not routinely accessible to scholars. This situation also limits the ability of researchers to study changes in diversion over time since intermittent data collection makes it impossible to routinely ascertain the level of variation in market indicators. The latter problem is partially resolved by the availability of firearm trace data, but this also suffers from intermittent analysis as well as the absence of information on ammunition.

To address these known limitations of the existing empirical literature, this chapter proposes and demonstrates the utility of examining ammunition found in and with firearms recovered by local law enforcement. These recovered firearms, the product of vehicle searches, pedestrian stops, 911 calls for a person with a firearm, as well as other modes of recovery, represent a window into the obtained supply of ammunition, and through it, the supply channels through which it was obtained. While recovered firearms have previously been used to measure firearm popularity as well as the impact of changes in weapon caliber on the probability of mortality (Braga and Cook 2018; Zimring 1972), this appears to be the first effort to examine ammunition loads rather than recovered weapons. Based on this novel and yet widely available source of information, we examine two measures of ammunition availability – firearm fullness and ammunition "scrounging." Firearm fullness, which we define in more detail below, is a measure of the degree to which firearms are fully or partially loaded. Ammunition "scrounging," by contrast, is a measure of the variety of manufacturers and batches observed in the loaded ammunition. Using data from the Boston Police Department to illustrate the utility of this information and these two measures, we show that most recovered firearms are substantially full, that the mode of recovery has seemingly little impact on the degree of fullness, that there is greater evidence of scrounging, and that firearm cost is uncorrelated with firearm fullness but is weakly correlated with ammunition scrounging. In addition, we find that many weapons are ready for immediate use due to the common practice of chambering rounds.

6.2 Prior Research on Illegal Firearm and Ammunition Markets

Guns and ammunition, like other manufactured goods, are distributed and sold through several distinct channels. They can be purchased from retail outlets, online stores, at gun shows and conventions, and they can be exchanged with or without financial remuneration. This constitutes the legal market for guns and ammunition. Unlike many other manufactured goods, however, they are also subject to distinctive regulatory restrictions that limit their purchase, ownership, possession, and use. These regulatory restrictions dictate who can legally purchase firearms, and to a lesser extent, who can legally purchase ammunition (Barragan et al. 2017). Prohibited groups include convicted felons, certain individuals experiencing acute mental health crises, as well as large numbers of children. While specifics vary from state to state, this basic framework is shared, and acts of regulatory compliance are carried out by the Bureau of Alcohol, Tobacco, and Firearms (ATF), which is a federal agency with oversight across all local jurisdictions, and many by local and state law enforcement agencies.

The exclusion of a population of potential consumers from the legal market for firearms and ammunition would not be enough to spawn an illegal market for firearms or for ammunition in the absence of some latent demand for firearms from within the excluded population. Decades of research and writing have repeatedly documented that there is enormous latent demand among these groups of excluded purchasers and possessors (Blumstein 1995; Cook 2000; Zimring 1972). And given this latent demand, scholars have spent the last 20 years trying to form a clearer picture of how legally manufactured and distributed firearms and ammunition move from the shelves of legal sellers into the hands of prohibited possessors and other excluded end-users. Through this work, they have identified several distinct supply channels, including fraudulent direct purchases from Federal Firearm Licensed (FFL) merchants, purchases at less closely monitored gun shows, and thefts from vehicles, homes, and businesses. Available evidence suggests that all of these routes play important roles in facilitating the acquisition and use of firearms by prohibited groups (Braga et al. 2012; Pierce et al. 2004; Sorenson and Vittes 2003). And while some regions show evidence of trafficking pipelines or other discrete point sources for the diversion of legal firearms, other jurisdictions appear to have much more diffuse diversion processes at work, with theft and secondary sales being the primary pathways through which guns are diverted (Braga et al. 2002, 2021; Decker, Pennell, and Caldwell 1995; Newbold 1999; Pierce et al. 2004).

Far less research has focused on the topic of ammunition. Studies have examined ammunition in Chicago, Los Angeles, the Netherlands, as well as the United Kingdom. These inquiries have simultaneously highlighted the duality of routine access to ammunition and at least some indicia of market thinness. For example, Tita and colleagues examined the records of FFL merchants in Los Angeles, California (Tita et al. 2006). Their analysis uncovered that 2.6% of purchasers at FFL merchants were prohibited by law from purchasing ammunition. These prohibited purchasers nonetheless purchased 10,050 rounds, or 2.3% of the rounds sold to all purchasers during the study period. A subsequent study in Los Angeles using interviews of jail inmates revealed that local retail establishments, especially those with substantial sporting goods sections (e.g., Big 5 Sporting Goods, Walmart), were the primary locations at which interviewees purchased ammunition (Chesnut et al. 2017). Respondents also reported that ammunition, at least in Los Angeles, was easier to obtain than firearms themselves. Corresponding work in the Netherlands has described a very similar situation with ammunition easily diverted from retail or sporting establishments (de Vries 2013).

By contrast, studies in Chicago have revealed a somewhat different picture of the illegal market for ammunition and firearms. Using interviews with market participants and trace data on recovered firearms, Cook et al. observed many indicators of market thinness for both firearms and ammunition (Cook et al. 2007). Their respondents reported large markups over retail prices for both, difficulty in buyers and sellers successfully transacting, and a generally fragile market for illegal weapons and ammunition. Their results even hinted at the possibility that the market for illegal ammunition in Chicago, unlike in Los Angeles, might be much thinner than the market for illegal firearms. These indicators of market thinness, including large mark-ups for ammunition, have also been reported in the United Kingdom (Hales, Lewis, and Silverstone 2006).

The reasons for this added possible thinness for ammunition relative to firearms are not well understood. It has been posited that there are fewer sellers or middlemen willing to sell ammunition. The added risk of providing unknown buyers with ammunition is also potentially physically risky for ammunition sellers in a way that selling an unloaded firearm is not. An additional explanation sometimes offered is that sellers of illegal firearms do not want to contribute to victimization within their communities and that they view the sale of firearms for self-protection as morally acceptable, but the sale of ammunition as morally questionable (Cook et al. 2007). Another way of reconciling the competing results from Los Angeles and Chicago (or the Netherlands and the United Kingdom, for that matter) is to say that illegal markets, just like legal markets, can vary. Los Angeles and the Netherlands, at least historically, both seem to share certain regulatory features that would make for relatively thick ammunition markets – plentiful supply, few regulatory hurdles, and the potential for prohibited possessors to access the legal market itself. Chicago and the United Kingdom, by contrast, have minimal supply, longstanding and substantial regulatory hurdles, and somewhat less direct legal market access. However, these are all just possible explanations for observed empirical differences.

To really understand why one setting might have a thinner or thicker illegal market for ammunition or an illegal market for ammunition that is thinner than the illegal market for firearms, it is necessary to have comparable data on illegal ammunition and firearms across settings. It would also be ideal to have this data across time, which would allow for a more empirical assessment of whether changes in the regulatory environment or enforcement pressure impact ammunition availability or price. While this form of routine information collection could be obtained from uniformly administered interviews of market participants or surveys of jail inmates, routine surveys of inmates have declined or refocused to other topics, and interviews of illegal market participants are necessarily few and far between. Instead, the collection and analysis of administrative data, if sufficiently substitutable, could overcome these problems.

Historically, administrative data on illegal firearm markets has been limited to information on firearms recovered by law enforcement. This information has allowed scholars to figure out the "time-to-crime" for trafficked weapons (Braga et al. 2012; Cook 2000), the popularity and lethality of different weapons (Zimring 1972), as well as other features of the illegal firearms market. At the same time, this information has not previously been used to study the market for illegal and/or diverted ammunition. This could be due to the lack of routine sharing of this information by law enforcement, which generally keeps information on recovered ammunition in their evidence recovery files rather than in their crime statistics databases. Another reason why this information may have been underutilized is the challenge of inferring features of the illegal ammunition market supply chain from recovered ammunition when it does not include classic market transaction information (e.g., price, mark-up, search costs). However, lurking within recovered ammunition is a surprising amount of information that speaks to the ease or difficulty with which market participants can obtain it. Consider, for the moment, how the preferences and constraints of market participants manifest within the ammunition found after shootings or in the firearms recovered after arrests.

If the market for ammunition is constrained and especially if it is highly constrained, we might expect to see shooting incidents with fairly low numbers of recovered shell casings, either because shooters lack the ability to obtain full magazines of ammunition, or, somewhat less plausibly, because shooters could be attempting to preserve at least some of their ammunition for future use. Likewise, recovered firearms, especially those not recovered in the context of a shooting incident, might be found partially full and sometimes completely empty, especially if the illegal market has entered a state of shortage. These examples illustrate the ways in which the needs of prohibited possessors could interact with the features of their supply markets to manifest in observable patterns of both recovered shell casings and recovered ammunition within recovered firearms.

To make sense of these possibilities, which we discuss with greater specificity in the following sections, we examine the fill levels for firearms recovered in the City of Boston. Boston serves as a particularly useful setting to investigate the supply of illegal ammunition because Massachusetts is one of just five states that have extended state firearm licensing requirements to the purchase of ammunition. Along with Connecticut, New Jersey, Illinois, and California, Massachusetts requires a firearm license to purchase or possess ammunition. This puts Massachusetts, and by extension Boston, at the far end of the distribution of localities with respect to regulatory scrutiny of would-be purchasers. For this reason, using it to illustrate the benefits of new metrics for ammunition diversion is doubly useful.

6.3 Data

The data for this study comes from a unique source of information on recovered firearms – social media.[1] The Boston Police Department (BPD), which is a large metropolitan police department, has for many years posted photographs and descriptions of most recovered firearms on the Department's social media channels to highlight the Department's ongoing efforts to combat gun violence. Unlike other, more occasional efforts, the Department reports posting nearly all recovered firearms with limited exclusions (i.e., homicides, domestic violence cases, as well as situations where the public affairs office is unable to track down the detective handling the case to get a photo).[2] This makes the posted photos to X (formerly Twitter), Facebook, and the Department's website a large repository of information on the condition of recovered firearms and their contents. Also unusual, compared to other police departments that post images of recovered firearms, is the quality of information posted. Most postings include not only a photo of the weapon but also a photo of the ammunition and ammunition storage devices recovered. This makes it possible to infer weapon manufacturer, caliber, maximum load level, observed load level, and ammunition mix. In addition, the textual description of the weapons and the circumstances of the arrest allow for further case details to be ascertained.

Each posted firearm from January 1, 2023, until September 18, 2023, was individually reviewed and coded. Information available from the visual images of recovered firearms included the condition of the firearm, the type of firearm (e.g., semi-automatic pistol, revolver, rifle), the capacity of the recovered magazine, the number of rounds recovered in the firearm, and the bullet type (e.g., conical, flat nosed, full metal jacketed, hollow point), as well as the condition of the recovered rounds. Information obtained from the textual description of the firearm recovery included the date of the recovery, the circumstances of the recovery (e.g., search warrant, traffic stop), the lead arrest charge (e.g., illegal possession of a firearm), the manufacturer of the firearm, and the model of the firearm.[3]

Given that the recovered firearms publicly shared via the internet represents a subset of all recovered firearms, which itself represents a subset of all illegal firearms in the City of Boston, one key empirical question is whether there is reason to believe that the probability of observation is sufficiently uncorrelated with the attributes of recovered firearms and recovered ammunition that any patterns observed in this sample is likely to adequately represent the full unobserved distribution. A second key empirical question is whether the information on ammunition count and quality contains a sufficiently clear signal that this can be used to assess market thinness and supply chain inventory in the absence of classic market indicators such as price, mark-up, as well as search costs. In the next section, we explore both questions in more depth.

6.4 Methods and Measures

Thin markets have few buyers and sellers. This leads to delays in completed transactions and price volatility when counterparties do transact. However, "thinness" is not just about how long it takes people to transact in quantitative terms or how well-priced assets are relative to some theoretical value. It is also about whether willing buyers and willing sellers transact and transact to the degree that they would like to. If buyers and sellers can successfully transact, but only after incurring large search costs, then the market is thin. If buyers and sellers can transact with or without large search costs, but supply is insufficient to meet demand, then the market is in shortage. Restating this for the case of illegally obtained ammunition, if buyers or end-users would like more ammunition than they have, then the market is thin. Whether due to price mark-ups or difficulty finding sellers, existing market transactions are insufficient to meet their latent demand. Conversely, if they have enough ammunition for their purposes, then the market can still be thin, but it is not so thin that market participants are effectively supply-constrained. This suggests that "thinness" in illegal ammunition markets can be assessed not only based on transaction records but also based on the amount of ammunition that interested buyers can obtain, since ammunition is the good being transacted and its presence or absence serves, when coupled with a clear signal of the amount of desired ammunition, as the clearest signal of whether all possible transactions have successfully occurred.

At a basic level, two principal ammunition metrics are available from firearms recovered by police: the number and type of rounds recovered.[4] However, these two measures by themselves are likely imperfect proxies for supply constraints and/or market thinness due to the lack of any information regarding the desired number or type of rounds. They become more useful when they are converted into two related metrics – the ammunition fill level and the firearm ammunition mix. The ammunition fill level, which we define as the number of rounds observed divided by the capacity of the weapon in which or with which they are recovered, offers a window into the accessibility of ammunition. If recovered firearms are mostly full, then willing buyers are able to obtain an adequate supply of ammunition. If recovered firearms are routinely observed to be partially filled in the absence of an indication that they have recently been fired, it can be reasoned that illegal firearm possessors have incomplete or insufficient access to ammunition. This necessarily tests a particular form of market thinness – that the price of ammunition is so high, or the availability is so low, that would-be weapon carriers are unable to obtain ammunition and are forced to carry an empty or substantially empty weapon. An empirical test, described in greater detail below, will assess how reasonable this test is.

Operationalizing this metric requires computing the fill level. For revolvers this is trivial. The number of chambers represents the maximum capacity of the firearm. For most semi-automated pistols, this is also relatively uncomplicated. Based on the specifications of the manufactured firearm and the observed magazine, it is possible to calculate how full the firearm is at the time of recovery. The existence of extended magazines or pistols that can accommodate different-sized magazines raises the question of whether a weapon is full when the number of observed rounds matches the smallest capacity of the firearm, or instead the full capacity of whichever magazine was recovered with the weapon itself. Similarly, there is a question of whether to include an additional chambered round in the full capacity of a semi-automatic pistol or not. Firearm manufacturers do not uniformly report the capacity of their firearms. Some exclude chambered rounds in their capacity specifications, while others include them, usually in the form of a number followed by +1, where the first number represents the round capacity of the magazine and the +1 represents the round that can be stored in the chamber. Since the calculation of fullness is designed to assist in measuring the difficulty of obtaining ammunition, not the degree of "readiness" of the firearm, we opted to consider a firearm as full if it contains a complete magazine, regardless of whether it also contains a chambered round. We similarly chose to consider the capacity of the magazine as the effective capacity of the firearm, even if the magazine is non-standard or extended, assuming that its size can be inferred. If it cannot be inferred, the capacity of the most common version of the firearm is used. This set of counting rules ensures that the theoretical construct at the heart of this study can be tested, as a greater fraction of recovered firearms are found to be fuller, the likelihood of market thinness goes down. Conversely, as a greater fraction of recovered firearms are found to be less full, the likelihood of market thinness goes up.

Ammunition mix represents a second metric. Like other goods that require consumable components to operate (e.g., razors/razor blades, cars/gas, flashlights/batteries), firearms require consumables to function. Multiple manufacturers' ammunition, due to standardization, can be used within a given firearm. However, unlike most consumer electronics, firearms can be operated even if they are incompletely loaded. In this respect, they are more like gasoline for cars. A car can still be driven with a partial tank, assuming the correct type of gasoline (e.g., diesel versus unleaded) has been added, regardless of brand (e.g., Exxon, Chevron). What this suggests is that features of the market for ammunition could manifest in the mix of ammunition observed within recovered firearms. If firearms are loaded with a single manufacturer's ammunition, then it was likely loaded from a single batch or box, indicating that the possessor was able to obtain a large enough supply to meet, at a minimum, his or her instantaneous needs for a loaded firearm. If firearms are loaded with more than one manufacturer's ammunition, then there are likely two different scenarios that could be occurring. In the first, an end user is refilling a partially full weapon from a new supply of ammunition, and this process is observed *in media res*. While the exact split observed could vary, the weapon should still be full (or full enough), and the ratio of one ammo type to the next should form a relatively even distribution. The other possibility is that more than two kinds of ammo are observed within a single recovered firearm. This almost certainly indicates that the firearm is not being reloaded from a standard source or supplier, which normally sells ammunition in boxes measured in dozens of bullets. A more cobbled together ammunition fill combining three or more distinct ammunition types suggests that the end user is scrounging for ammunition across multiple sources, any one of which can only offer a single or a few bullets, or that the end user is receiving from a supplier who has aggregated from a similarly defuse set of sources such that they cannot offer a single type of ammunition. Given that the act of aggregation, whether by the end user or a middleman, necessarily involves additional effort, it seems reasonable to assume that evidence of aggregation should not be interpreted as a taste preference. Instead, it is more logical to view this as evidence of a constraint that limits aggregators from accessing more conventional supplies of ammunition.

We now return to the potential problem of the unrepresentativeness of recovered firearms as stand-ins for all firearms. Since firearm end-users choose when to venture out of their homes with their illegal firearms, one could imagine a situation in which only those with access to sufficient ammunition would dare to venture out with their weapons, thereby creating an empirical distribution that is a poor proxy for the overall distribution of end-users. This potential problem would be especially difficult if all firearms were recovered via the same search strategy by police. For example, if all or nearly all firearms were recovered via pedestrian stops or car stops. However, this is not the case. Firearms are routinely recovered through a variety of different police search strategies. Pedestrian and car stops are two, but warrant searches, calls of an armed gunman, shooting reports, and drug investigations are also common routes to a recovered firearm. This diversity of routes loosely corresponds to the multiple sampling strategies used in sampling other hidden (Chan, Silverman, and Vincent 2021). For our purposes, multiple recovery strategies provide an opportunity to test whether the route of recovery predicts the observed distribution of either fullness or ammunition mix (alongside firearm type, caliber, etc.).

This range of circumstances leading to the recovery of illegal firearms can usefully be grouped into three distinct modes of recovery. First, firearms can be recovered after a shooting or after a report of a person with a gun. These situations almost always occur in public and most often involve the visible display and sometimes discharge of the firearm. Second, firearms can be recovered from a person after a traffic stop, a pedestrian stop, or some other police search of an individual observed in public. Finally, firearms can be recovered from a private residence after the execution of a search warrant, often involving a drug investigation, or after responding to a call of domestic violence. The assumption of our sensitivity test is that any bias in our estimates due to a positive correlation between ammunition load level and probability of recovery will most likely be observed in our first two categories, since these weapons are being carried into public spaces for potential or actual use. By contrast, we assume that weapons recovered in private dwellings in searches that should not be temporally correlated with the load status of the recovered weapons can serve as an empirical check on the degree of any induced correlation in our estimates.

To test this, we fit a regression model predicting the probability of a firearm being full on the mode of recovery, weapon characteristics, and ammunition mix. Doing so also provides an opportunity to test for other potential observable threats to the validity of inference. For example, firearms recovered immediately after a reported shooting should be observed to have lower fullness levels, albeit unrelated to market thinness. Weapons involving higher calibers, which cost more to load, should be more likely to be empty or partially full. We explore these and related questions in our next section.

6.5 Results

6.5.1 Ammunition Levels

Table 6.1 reports the basic level of fullness for the firearms recovered by the BPD between January and September of 2023. Of the 194 weapons studied, 90% had sufficient information available that fullness could be inferred. The remaining 10% may or may not be missing at random, and additional analyses of these weapons are ongoing. Just under 50% of firearms (48.5%) were found to be fully loaded. Another 18% were found to be between 80% and 99% loaded when recovered. Fifteen percent were found to be between 50% and 79% loaded. And just under 8% were found to be less than 50% loaded. Of these, just 1% were found to be empty. Figure 6.1 reports the same results graphically, but the exclusion of missing values shrinks the denominator and leads to slightly higher estimates of remaining categories. For example, the percentage of fully loaded firearms goes from 48% to 54% of cases. Regardless of how missing values are handled, the overall estimate is similar – approximately half of all recovered firearms are full, most non-full firearms are substantially full, and a minority of firearms are less than half full.

TABLE 6.1

Fullness of Recovered Firearms

Fullness	Percentage (%)	*N*
100%	48.45	94
0–99%	18.04	35
50–79%	15.46	30
01–49%	06.70	13
Empty	01.03	02
Missing fullness information	10.31	20
N		194

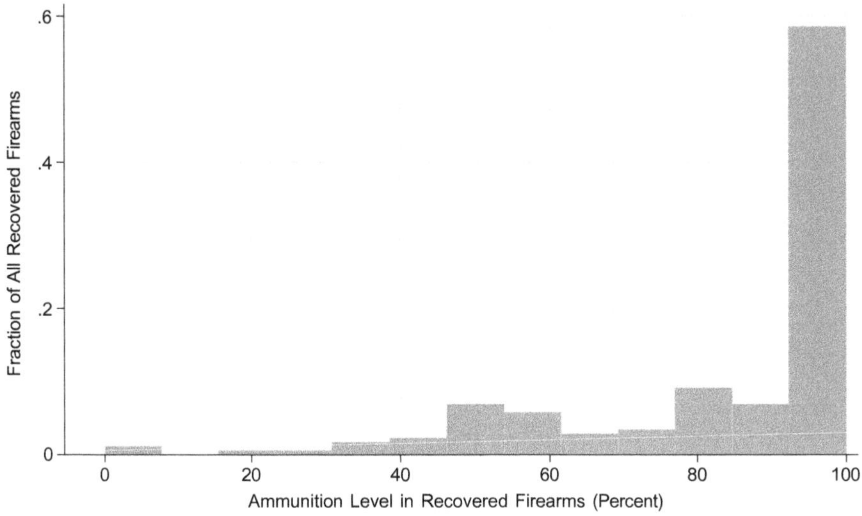

FIGURE 6.1
Distribution of Ammunition Levels Found in Recovered Firearms.

6.5.2 Ammunition Mix

Table 6.2 reports the basic results on the mix of ammunition found in recovered firearms. Due to the way in which BPD photographed ammunition, the data missingness rises to 26% of firearms. Still, the results for non-missing cases are intriguing and show a less skewed distribution than was observed for ammunition levels. Thirty-three percent of firearms were found to have a single type of ammunition regardless of fill level. Another 19% were found to have two distinct types of ammunition. And 22% of recovered firearms were found to have three or more distinct types of ammunition. This reduced level of skewness across the ammunition mix metric hints at the possibility, discussed earlier in the paper, that the two measures of market thinness in this paper could diverge from each other in important respects.

Figure 6.2 graphically explores this possibility by re-graphing Figure 6.1 over the three ammunition mix categories. While small differences, some of which could be statistically significant in a larger sample, can be seen, the overall shape of the distributions is largely the same across categories. It also closely resembles the distribution seen in Figure 6.1. Whether prohibited possessors are scrounging, loading, or doing something in between (e.g., scrounging or reloading), the probability of their having a full or nearly full firearm is quite comparable. This suggests that these metrics give different impressions of market thinness due to their lack of obvious co-variation. One further example of a complex relationship between different measures of ammunition can be seen in the relationship between ammunition quality and ammunition mix or scrounging. In additional results (not shown), it was

TABLE 6.2

Number of Distinct Ammunition Varieties in Recovered Firearms

# of Observed Ammo Types	Percentage (%)	*N*
1	32.99	65
2	19.07	40
3 or more	22.16	44
Missing type information	25.77	55
N		194

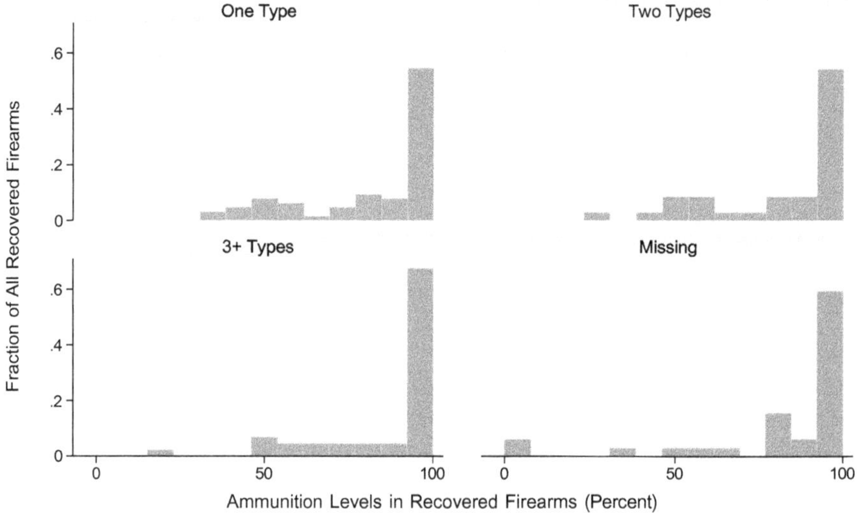

FIGURE 6.2
Distribution of Ammunition Levels Found in Recovered Firearms by Number of Distinct Ammunition Types

observed that not all cases of ammunition scrounging (3+) included visually low-quality ammunition. However, most low-quality ammunition was observed in cases involving scrounging.

6.5.3 Sensitivity Check

Figure 6.3 serves as a preliminary and visual check of the sensitivity of our primary measure – ammunition level – to concerns of omitted variable bias. If the distribution of ammunition load levels differs notably depending on the mode of firearm recovery (e.g., recovered after a shooting, recovered after a traffic stop, recovered after a warrant search of a dwelling), then the validity of this as a measure of thinness in the market for illegal ammunition would be called into question. However, Figure 6.2 reveals that load level distributions look generally quite similar across modes of recovery. Weapons recovered in private residences appear to have very similar load distributions as weapons recovered in public places. Weapons recovered after the report of a shooting or a public display of a firearm are the only categories that hint at a deviation from the overall distribution for recovered firearms. However, the deviation is consistent with the hypothesis that firearms recovered through this recovery pathway should have at least a slightly lower probability of being found full. Therefore, Table 6.4 tends to confirm that the mode of recovery has a minimal relationship with the probability of recovering a full firearm.

Table 6.3 reports the results of our formal test of whether ammunition levels varied with mode of recovery. Because of the extreme skewness of the distribution of the outcome measure, as seen in Figure 6.1, we elected to use a limited dependent variable model to represent the outcome. This model predicted the probability of a recovered firearm being full on the indicators for the three modes of recovery, with recovered from a traffic/pedestrian stop being the holdout category. Additional indicators for the number of batches and the weapon manufacturer were also added to this regression model. The results reveal that there were no statistically significant differences between the modes of recovery, confirming the graphical results seen in Figure 6.3. Similarly, the results for the ammunition mix metric also show no significant differences, confirming the graphical results seen in Figure 6.3. Interestingly, no clear pattern of coefficients or significance levels appeared for manufacturers either.[5] Given the sizable price differentials between a good condition

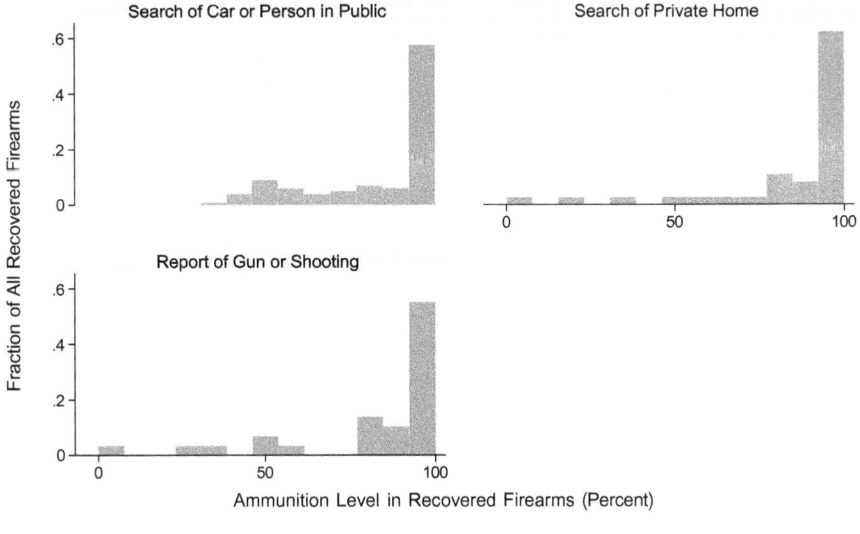

FIGURE 6.3
Distribution of Ammunition Levels Found in Recovered Firearms by Circumstances of Recovery.

TABLE 6.3

Logistic Regression of Binarized Fullness of Recovered Firearms on Select Features of Firearms and Circumstances of Recovery

	Coefficients	S.E.	*P* Values
Mode of Recovery			
Search of a private residence	1.18	0.524	0.708
Search of a car or person in public	1.14	0.589	0.806
Ammunition Mix			
Two types	0.833	0.382	0.690
Three or more types	1.290	0.549	0.560
Missing type info	1.081	0.509	0.868
Manufacturer			
Smith & Wesson	0.611	0.325	0.354
Taurus	1.077	0.605	0.895
Ruger	0.515	0.332	0.562
P80/ghost gun	0.515	0.332	0.303
Other	0.520	0.219	0.228
Constant	1.323	0.707	0.120
$N = 165$			

Notes: The reference category for Mode of Recovery was a report of a gun or shooting. The reference category for Ammunition Mix was a single type of ammunition. The reference category for manufacturer was Glock.

Glock or Smith & Wesson compared to a Taurus, it appears, based on this sample, that weapon price and ammunition level are not well-correlated. However, some of these results could change with a larger sample size (see Figure 6.4 for details).

6.5.4 Inferring Firearm Use Intent

It is common to encounter a range of public and scholarly narratives regarding the intent of illegal gun possessors. Some clearly emphasize that illegal possessors carry their firearms for aggressive purposes – to seek revenge or otherwise enact violence, a view consistent with the observed levels of actual gun violence. Other narratives indicate that illegal possessors often carry illegal firearms for purely defensive purposes and would only brandish or discharge their firearms if first threatened (Decker, Pennell, and Caldwell 1995; Watkins, Huebner, and Decker 2008). Still others admit to aggressive uses (e.g., brandishing or threatening) but distinguish their more "virtuous"

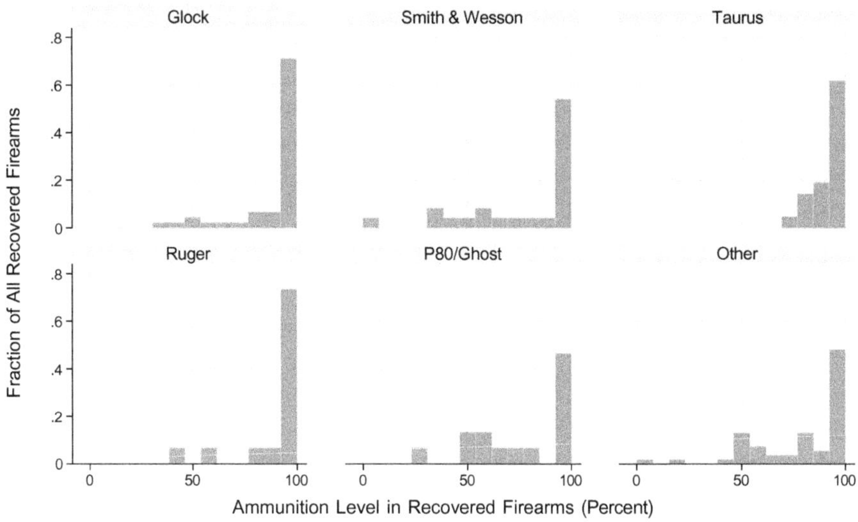

FIGURE 6.4
Distribution of Ammunition Levels Found in Recovered Firearms by Firearm Manufacturer.

illegal possessions from more problematic ones by stating that their weapons are carried empty. As one such individual reported in a recent ethnography,

> I'm stealing a lot of car radios right now, and sometimes, if I get really brave, I may try to take a purse. For that shit, I keep the gun, but I never use it, you know. I don't even load it, I keep the bullets I got for the bigger shit I do.
>
> **(Cook et al. 2007, F598)**

In the past, the only way of measuring the prevalence of these different states of intent was to ask respondents using self-report surveys or interviews. However, there is some reason to be concerned that the concept of defense carrying, which is reported by most respondents in self-report surveys, is being over-reported since it also happens to be the socially acceptable response (Chesnut et al. 2017; Dong and Wiebe 2018; Lizotte et al. 2000; Watkins, Huebner, and Decker 2008). Theoretically, criminal history searches could also be used to establish how often individuals arrested for illegal firearm possession had previously or subsequently been arrested for illegal discharge of a firearm, as another way of making sense of how often firearm possession is a reasonable proxy for firearm use intent. However, with homicide clearance rates at or below 50% for many cities and non-fatal shooting rates substantially lower, it's not clear how well official records capture the true extent of firearm violence involvement among firearm possession arrestee populations (Cook et al. 2019).

Here again, recovered ammunition offers a window into something that previously was partially visible. The relevant information comes in two forms. First, within the larger distribution of ammunition levels observed in recovered firearms is the very particular number of zero, reflecting those situations in which police recovered an unloaded firearm. This can either be because the firearm was empty, and the illegal possessor lacked ammunition, or because the firearm was empty, and the illegal possessor was found to have ammunition separately in his or her possession. The frequency with which either of these situations occurs, but especially the first situation, serves as a reasonable and direct measure of possession without the practical intent of harming through firearm discharge if it is assumed that an empty firearm is known and intended to be empty by the possessor.[6] Table 6.1 previously reported that of the 194 recovered firearms in this study, roughly 1% were found to be empty. This suggests that empty weapon possession is not well-established by examining recovered firearms. However, a closer examination of the two cases that contribute to the 1% estimate reveals an even starker picture. In the first of these two cases, the firearm, a revolver, was recovered through service of a warrant in a drug investigation, and though empty, 32 rounds of matching .22 caliber ammunition were also recovered. In the other case, the firearm, a Smith & Wesson, was recovered empty after police received a report of a person with a gun, which could indicate that a shooting had taken place. Whether the exact estimate from this sample is rendered as 1%, 0.05%, or 0%, the implication is the same – illegal firearm possessors in Boston don't normally possess empty firearms.

We next examined the frequency with which recovered firearms had a round chambered. This is observed in just under half (46%) of recovered firearms. However, this result ignores that while most recovered firearms are semi-automatic pistols, in which a chambered round no longer requires racking of the slide prior to discharge of the weapon, a small number of recovered firearms were revolvers, which, when fully loaded, also have a round chambered. Including the four fully loaded revolvers with the existing count of chambered firearms bumps up the observed percentage of chambered firearms to 48% of all weapons.

Figure 6.5 explores chambering behavior in more detail by comparing the probability of chambering to the overall ammunition level in recovered firearms. If chambering is simply reflective of possessors' desire to carry their weapons in a state of maximum fullness, then there should be little chambering at lower levels of fullness. Conversely, if chambering is observed across levels of fullness, then the decision to chamber a round is likely more reflective of intent to have firearms "ready-to-go." Figure 6.5 shows that there is a weak but positive relationship between ammunition level and probability of chambering.[7] To remove any motive contamination from our estimate of chambering for "ready-to-go" purposes, we then computed the percent of weapon chambered among full and partially full firearms separately. The full firearms were observed to have a 50% probability of having a round

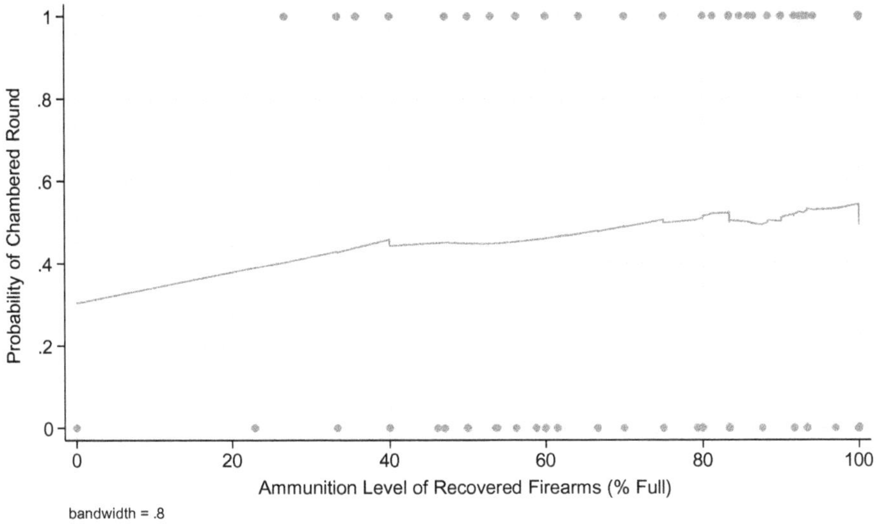

FIGURE 6.5
Lowess Smoother Estimate of Round Chambering by Ammunition Levels.

chambered, reflecting the fact that our definition of "full" was restricted to full magazines only. All other firearms were observed to have a 43% probability of having a round chambered. The similarity of these two estimates, which could disappear or gain statistical significance in a larger sample, for now suggests that many illegal possessors are interested in having their weapons "ready to go," not simply that they want to make sure that they have maximum ammunition loads.

6.6 Discussion

Scholars have long sought to better understand the dynamics of the illegal market for firearms and ammunition. And by using interviews of market participants, firearm tracing, and surveys of jail inmates, the literature on illegal firearm markets is now rich with hypothesized mechanisms, variables, nuances, and narratives. However, determining which of these are the dominant features of the illegal market for ammunition and which of these are comparatively small players in the informal market through which guns and ammunition transition from store shelves to the streets remains an ongoing and well-recognized problem. The present investigation has proposed one often-overlooked source of information (namely recovered firearms and ammunition) that could shed new and potentially inexpensive light on the supply channel for ammunition. This information is available retrospectively and prospectively, in small and large jurisdictions, as well as potentially being already digitized for jurisdictions that have digitized their evidence-recovery systems.

Using this information for one major metropolitan jurisdiction, we have shown that the market for illegal ammunition, at least in this jurisdiction, is only somewhat thin. While many weapons have a mixture of multiple types of ammunition, roughly 50% of weapons are recovered with full magazines, and another sizable portion are recovered substantially full. This pattern might be interpreted more optimistically if these results were observed in a jurisdiction with a weaker ammunition regulatory regime in place, but the research setting happens to have one of the most restrictive regulatory regimes in the country. This suggests that access to ammunition is a surmountable problem. This result should not be interpreted as indicating the fruitlessness of attempting to interrupt the diversion of ammunition into the illegal firearms and ammunition market. However, it should temper our understanding of the effectiveness of existing measures. At a theoretical level, it should also temper our enthusiasm for the attractiveness of theories of illegal markets that emphasize their distinctiveness relative to more conventional markets. Minors routinely obtain cigarettes and alcohol. Prohibited possessors of firearms, a decent subset of whom are minors, routinely obtain firearms. Why should ammunition be any different?

A separate point also bears mentioning. An ongoing theoretical and policy debate has focused on the degree to which these same prohibited possessors, again, many of them juveniles, are carrying firearms defensively due to fears for their personal safety. Examples of this can be seen in debates on the proper sentences for gun possession cases (Friedersdorf 2022). This is a larger discussion than this narrow analysis can address. However, on one sub-point of this larger debate, the finding of this descriptive analysis is quite clear. It has been posited that some fraction of prohibited possessors carry unloaded or minimally loaded firearms to avoid injuring others. These unloaded but carried firearms, so the theory goes, can be used to deter would-be assailants, pacify robbery victims without risk of accidental discharge, or otherwise "safely" avoid actual violence through the possession of a weapon that can be visually confused for one with the real potential for violence. While this may be true somewhere at some time, it is disconfirmed by what has been observed in this study.

Still, the results reported in this initial empirical inquiry should not be over-interpreted. This is a study of only 7.5 months of data for one city. Additional work to lower the observed level of missingness remains to be completed, and having a larger sample size would likely shift some of the observed and statistically insignificant relationships into significance. Furthermore, despite the sensitivity checks reported in this chapter, it is possible that recovered firearms are an imperfect proxy for unrecovered illegal firearms. Finally, firearm fullness, while useful as an instantaneous measure of ammunition accessibility, provides little insight into other aspects of the illegal ammunition market. For example, it provides little insight into price and search costs. Both of which have been speculated about in the qualitative literature but have remained unexamined in a more systematic fashion.

In future work, beyond replicating the results reported in this chapter with larger samples and different police agencies, there is a large range of additional inquiries that could be pursued with this administrative data. For example, studies could be conducted comparing ammunition levels in cities with a greater number of "community guns" (e.g., New York City), gang-controlled guns (e.g., Chicago), as well as individually obtained firearms (e.g., Washington, D.C.). Likewise, studies could compare ammunition levels in firearms linked to shootings and those that are not to examine whether there is a split market for ammunition between violence-connected and violence-unconnected firearms. Apart from comparing jurisdictions based on features of their gun violence patterns or their regulatory regimes, the measures reported in this chapter could also be used to explore the impact of specific enforcement actions at relatively low cost. Future work could also usefully revisit three aspects of this analysis – ammunition condition, caliber, and manufacturer/price.

The present analysis treated the visual condition of the recovered ammunition as the object of interest. However, several other visual attributes of ammunition could also have been explored. For example, brass ammunition is generally more expensive than steel ammunition. Examining the pattern of brass versus steel ammunition would therefore provide a window into the resources of the purchasers. It may also be possible to ascertain whether ammunition was factory-loaded or reloaded. Likewise, future work could usefully analyze the "commonness" of the chambering of the recovered firearms. Unusual chambering generally increases the cost of ammunition and the difficulty of obtaining it. Finally, while the present investigation focused on the firearm manufacturer as a proxy for the cost of the firearm (e.g., Taurus versus Glock), the retail prices of these weapons could also be directly calculated. All of these would further add to our understanding of the dynamics of illegal markets for firearms and ammunition.

Acknowledgements

The author would like to thank John MacDonald (University of Pennsylvania), Anthony Braga (University of Pennsylvania), Mark Coggeshall (CSOSA), and Peter Reuter (University of Maryland) for helpful discussions that culminated in this chapter. He would also like to thank Brian Wade (University of Pennsylvania), Steve Raphael (UC Berkeley), and seminar participants at Cal Berkeley and the NBER Workshop on the Economics of Firearm Markets, Crime, and Gun Violence (2024) for feedback on an earlier version of this manuscript.

Notes

1. BPD firearm postings were originally located on the social media website formerly known as Twitter. The full sample, however, was downloaded from the Boston Police Department's own news website – https://police.boston.gov/index.php. It is also available through the Department's Facebook page.
2. Author's personal communication with Boston Police Department May 26, 2023.
3. Not all textual descriptions contained manufacturer and model information. For these cases, as well as most other cases, each weapon was checked against the manufacturer's website to determine whether the manufacturer and model information was correct.
4. A third feature, condition or quality of rounds recovered, is also available.
5. The low representation of low-quality or inexpensive "junk" guns in this sample mirrors previously reported findings examining recovered firearms in Boston (Hureau and Braga 2018).
6. Empty firearms, in this interpretation, are treated as choices rather than as undesired or less desired outcomes of a constrained decision process. While this assumption is different than the previous assumption that empty firearms are a signal of highly constrained choices, the practical significance of this alternative interpretation is likely to be quite limited. The applicability of each theory can be most clearly seen in whether any empty firearms are recovered with accompanying but unloaded ammunition or if they are recovered without ammunition at all.
7. These two possibilities represent ideal types. In practice, there are several practical reasons why the empirical distribution is unlikely to perfectly conform to these patterns. For example, many illegal firearm possessors carry their weapons unholstered. This fact could easily generate a counter-pressure to not keep their weapons "ready-to-go" for fear of an accidental discharge. Similarly, removing a chambered round from a semi-automatic pistol is not as straightforward an exercise as removing a magazine from a pistol. Both of these considerations could depress the observed level of chambering.

References

Barragan, Melissa, Kelsie Y. Chesnut, Jason Gravel, Natalie A. Pifer, Keramet Reiter, Nicole Sherman, and George Tita. 2017. "Prohibited Possessors and the Law: How Inmates in Los Angeles Jails Understand Firearm and Ammunition Regulations." *RSF* 3 (5): 141–63. https://doi.org/10.7758/RSF.2017.3.5.07.Blumstein, Alfred. 1995. "Youth Violence, Guns, and the Illicit-Drug Industry." *The Journal of Criminal Law and Criminology (1973-)* 86 (1): 10–36. https://doi.org/10.2307/1143998.

Braga, Anthony A., Rod K. Brunson, Philip J. Cook, Brandon Turchan, and Brian Wade. 2021. "Underground Gun Markets and the Flow of Illegal Guns into the Bronx and Brooklyn: A Mixed Methods Analysis." *Journal of Urban Health: Bulletin of the New York Academy of Medicine* 98 (5): 596–608. https://doi.org/10.1007/s11524-020-00477-z.

Braga, Anthony A., and Philip J. Cook. 2018. "The Association of Firearm Caliber with Likelihood of Death From Gunshot Injury in Criminal Assaults." *JAMA Network Open* 1 (3): e180833. https://doi.org/10.1001/jamanetworkopen.2018.0833.

Braga, Anthony A., Philip J. Cook, David M. Kennedy, and Mark H. Moore. 2002. "The Illegal Supply of Firearms." *Crime and Justice* 29 (January): 319–52. https://doi.org/10.1086/652223.

Braga, Anthony A., Garen J. Wintemute, Glenn L. Pierce, Philip J. Cook, and Greg Ridgeway. 2012. "Interpreting the Empirical Evidence on Illegal Gun Market Dynamics." *Journal of Urban Health* 89 (5): 779–93. https://doi.org/10.1007/s11524-012-9681-y.

Chan, Lax, Bernard W. Silverman, and Kyle Vincent. 2021. "Multiple Systems Estimation for Sparse Capture Data: Inferential Challenges When There Are Nonoverlapping Lists." *Journal of the American Statistical Association* 116 (535): 1297–306. https://doi.org/10.1080/01621459.2019.1708748.

Chesnut, Kelsie Y., Melissa Barragan, Jason Gravel, Natalie A. Pifer, Keramet Reiter, Nicole Sherman, and George E. Tita. 2017. "Not an 'Iron Pipeline', but Many Capillaries: Regulating Passive Transactions in Los Angeles' Secondary, Illegal Gun Market." *Injury Prevention* 23 (4): 226–31. https://doi.org/10.1136/injuryprev-2016-042088.

Cook, Philip J. 2000. *Gun Violence: The Real Costs.* Studies in Crime and Public Policy. Oxford; New York: Oxford University Press.

Cook, Philip J., Anthony A. Braga, Brandon S. Turchan, and Lisa M. Barao. 2019. "Why Do Gun Murders Have a Higher Clearance Rate than Gunshot Assaults?" *Criminology & Public Policy* 18 (3): 525–51. https://doi.org/10.1111/1745-9133.12451.

Cook, Philip J., Jens Ludwig, Sudhir Venkatesh, and Anthony A. Braga. 2007. "Underground Gun Markets." *The Economic Journal* 117 (524): F588–618.

Decker, Scott H., Susan Pennell, and Amy Caldwell. 1995. "Arrestees and Guns: Monitoring the Illegal Firearms Market." *National Institute of Justice Research Preview.* Washington, District of Columbia: US Department of Justice. https://www.ojp.gov/library/publications/arrestees-and-guns-monitoring-illegal-firearms-market.

Dong, Beidi, and Douglas J. Wiebe. 2018. "Violence and beyond: Life-Course Features of Handgun Carrying in the Urban United States and the Associated Long-Term Life Consequences." *Journal of Criminal Justice* 54 (January): 1–11. https://doi.org/10.1016/j.jcrimjus.2017.11.002.

Durlauf, Steven, and Lawrence E. Blume. 2016. *The New Palgrave Dictionary of Economics.* Springer. https://books.google.com/books?hl=en&lr=&id=EO40DAAAQBAJ&oi=fnd&pg=PR2&dq=palgraves+economics+dictionary&ots=8ldaIVAGl-&sig=sT6z9N8nirOzPUrsPGWZa-RRQwg.

Friedersdorf, Conor. 2022. "The Anti-Gun Laws That Make Progressives Uneasy." *The Atlantic* (blog). February 10, 2022. https://www.theatlantic.com/ideas/archive/2022/02/blue-americas-new-gun-control-debate/622035/.

Gramlich, John. 2023. "What the Data Says about Gun Deaths in the U.S." *Pew Research Center* (blog). 2023. https://www.pewresearch.org/short-reads/2023/04/26/what-the-data-says-about-gun-deaths-in-the-u-s/.

GVA. 2023. "Gun Violence Archive." 2023. https://www.gunviolencearchive.org/.

Hales, Gavin, Chris Lewis, and Daniel Silverstone. 2006. "Gun Crime: The Market in and Use of Illegal Firearms." London: Home Office.

Harlow, Caroline Wolf, Bureau of Justice Statistics (BJS), US Dept of Justice, Office of Justice Programs, and United States of America. 2001. "Firearms Use by Offenders: 1997 Survey of Inmates in State and Federal Correctional Facilities." *Bureau of Justice Statistics Special Report, NCJ* 189369. https://www.ojp.gov/library/publications/firearm-use-offenders-1997-survey-inmates-state-and-federal-correctional.

Hayenga, M. L., A. B. Gardner, and J. P. Houck. 1978. "The Concept of a Thin Market." 261. North Central Regional Research Publication.

Hureau, David M., and Anthony A. Braga. 2018. "The Trade in Tools: The Market for Illicit Guns in High-Risk Networks." *Criminology* 56 (3): 510–45. https://doi.org/10.1111/1745-9125.12187.

Kena, Grace, and Jennifer L. Truman. 2022. *Trends and Patterns in Firearm Violence, 1993–2018*. US Department of Justice, Office of Justice Programs, Bureau of Justice …. https://bjs.ojp.gov/sites/g/files/xyckuh236/files/media/document/tpfv9318.pdf.

Lehmann, Fritz. 1938. "Thin Markets." *Social Research* 5 (1): 55–69.

Lizotte, Alan J., Marvin D. Krohn, James C. Howell, Kimberly Tobin, and Gregory J. Howard. 2000. "Factors Influencing Gun Carrying Among Young Urban Males Over the Adolescent-Young Adult Life Course*." *Criminology* 38 (3): 811–34. https://doi.org/10.1111/j.1745-9125.2000.tb00907.x.

Newbold, Greg. 1999. "The Criminal Use of Firearms in New Zealand." *Australian & New Zealand Journal of Criminology* 32 (1): 61–78. https://doi.org/10.1177/000486589903200106.

Pierce, Glenn L., Anthony A. Braga, Raymond R. Hayatt, and Christopher S. Koper. 2004. "Characteristics and Dynamics of Illegal Firearms Markets: Implications for a Supply-Side Enforcement Strategy." *Justice Quarterly* 21 (2): 391–422. https://doi.org/10.1080/07418820400095851.

Sheley, Joseph F., and James D. Wright. 1993. *Gun Acquisition and Possession in Selected Juvenile Samples*. US Department of Justice, Office of Justice Programs, National Institute of …. https://books.google.com/books?hl=en&lr=&id=CZzaAAAAMAAJ&oi=fnd&dq=gun+a cquisition+and+possession+in+selected&ots=XOTCgVyWHc&sig=EVDsHgZ5Fu8Kvk HQOVgQaBV5V9c.

Sorenson, Susan B., and Katherine A. Vittes. 2003. "Buying a Handgun for Someone Else: Firearm Dealer Willingness to Sell." *Injury Prevention* 9 (2): 147–50.

Terry, Don. 1992. "How Criminals Get Their Guns: In Short, All Too Easily." *The New York Times*, March 11, 1992, sec. A.

Tita, George E., Anthony A. Braga, Greg Ridgeway, and Glenn L. Pierce. 2006. "The Criminal Purchase of Firearm Ammunition." *Injury Prevention* 12 (5): 308–11.

Vries, Marsha S. de. 2013. "From the Instrument of Delivery to the Actual Agent of Harm: Fighting the Criminal Purchase of Ammunition." *European Journal on Criminal Policy and Research* 19 (1): 1–14. https://doi.org/10.1007/s10610-012-9173-3.

Watkins, Adam M., Beth M. Huebner, and Scott H. Decker. 2008. "Patterns of Gun Acquisition, Carrying, and Use Among Juvenile and Adult Arrestees: Evidence from a High-Crime City." *Justice Quarterly* 25 (4): 674–700. https://doi.org/10.1080/07418820801930118.

Zimring, Franklin E. 1972. "The Medium Is the Message: Firearm Caliber as a Determinant of Death from Assault." *The Journal of Legal Studies* 1 (1): 97–123.

7

Statistical Methods to Estimate the Impact of Gun Policy on Gun Violence

Eli Ben-Michael, Mitchell L. Doucette, Alexander D. McCourt, Avi Feller, and Elizabeth A. Stuart

Gun violence is a critical public health and safety concern in the United States. There is considerable variability in policy proposals meant to curb gun violence, ranging from increasing gun availability to deter potential assailants (e.g., concealed carry laws or arming school teachers) to restricting access to firearms (e.g., universal background checks or banning assault weapons). Many studies quantify policy impacts on gun violence using methods that combine state-level variation in the enactment of these policies with repeated measurements of the outcome of interest both before and after the policy change. In this chapter, we highlight the key features of the design and analysis of such evaluations, along with the associated opportunities and challenges, especially for studying gun policies in particular. We focus primarily on difference-in-differences and synthetic controls methods when there is staggered adoption of policies across jurisdictions, estimating the impacts of right-to-carry laws on violent crime as a case study.

7.1 Introduction

Rigorous evaluations of public policy are of the utmost importance. Understanding whether a policy has had its intended effect or an unintended one is essential for making future policy decisions. This is especially true for firearm policy. The United States has the highest rate of firearm violence among developed nations, with a gun homicide rate roughly 25 times higher than other high-income countries (Bangalore and Messerli, 2013). Firearm violence is also the leading cause of death for American adolescents under the age of 21, and firearm homicides disproportionately impact young Black Americans (Davis et al., 2023). Over the past several decades, lawmakers have attempted to address gun violence in many ways, but the most

 DOI: 10.1201/9781003430773-8

prominent approach has been laws that regulate the purchase, possession, and use of guns.

Understanding the impact of such policies is important but challenging, and researchers must overcome several significant hurdles. The research has been severely underfunded overall and relative to other leading causes of death (Ault, 2021; Stark and Shah, 2017). Data availability has been a barrier for researchers: analyses of gun policies are hindered by severe limitations on the type and scope of data available (Rowhani-Rahbar et al., 2019). There have also been challenges related to statistical methodology: many researchers seek to use variation in state policies to draw causal inferences, but reliable study designs are often difficult to craft. Despite these challenges, gun policy studies are quite impactful. Policy makers and advocates use these studies to push for adoption of new, evidence-based gun policies. For these reasons, it is essential that gun policy researchers use rigorous methods to estimate the effects of these laws.

In this chapter we review emerging methodological best practices for policy evaluation using "panel data" consisting of repeated measurements of the outcome of interest over time measured on relevant geographic units (such as states), with individual states adopting or removing the policy of interest at different times. These methods take advantage of longitudinal data to compare outcomes in jurisdictions that do and do not implement the policy, while adjusting for baseline differences between the jurisdictions in the pre-policy period to disentangle the policy's effects from other differences between jurisdictions. This chapter highlights the key features of the design and analysis of such evaluations along with associated opportunities and challenges, especially for studying firearm policies in particular. The goal is to help researchers construct trustworthy studies and to help consumers of research understand which results are trustworthy.

As a guide, we will use the notion of *target trial emulation* (Danaei et al., 2018; Dickerman et al., 2019), where we "design" an observational study as we would a randomized one, clearly defining the sample, including inclusion/exclusion criteria and exposure definition, outcome measurement, and the timing of all variables. Applying these concepts to policy evaluation with panel data gives *policy trial emulation* (Ben-Michael et al., 2021b), a design-focused framework that nests existing approaches and foregrounds the fundamental analytical decisions. As we discuss below, these ideas are in contrast to traditional approaches for impact evaluation with panel data that rely heavily on parametric regression models, such as two-way fixed effects approaches, that do not have a clear design aspect. Rather, these approaches have many drawbacks and can sometimes lead to biased results (Goodman-Bacon, 2021).

7.1.1 Firearm Policies and Impact Evaluations

Most gun laws in the United States are enacted and implemented at the state level, and historically states have approached gun policy in varied ways, resulting in a patchwork of different regulations. Writing in dissent in *New State Ice Co. v. Liebmann*, 285 U.S. 262 (1932), Justice Louis Brandeis argued that such policy variation creates a "laboratory of democracy." Indeed, this creates an environment that is ripe for causal inference – at least in principle.

State gun laws typically seek to regulate who can own or acquire guns, including the purchase process, possession, or use of a gun, or other behaviors related to ownership of guns. State laws typically list categories of individuals who are prohibited from possessing guns (e.g., those convicted of felonies, those with a specific history of alcohol misuse, those convicted of certain juvenile offenses, etc.) (Smart et al., 2023). Laws regulating the purchase process typically involve a background check that seeks to determine whether a prospective purchaser is in one of these prohibited categories (Vernick et al., 2017). As of August 2024, 28 states do not require a background check for private transfers (those not involving a licensed firearms dealer), but 22 states and the District of Columbia require a background check as part of the purchase or acquisition process. These background checks may occur at the point-of-sale (17 states + DC) or may be implemented as part of a permit or license process (9 states + DC) (Vernick et al., 2017; McCourt et al., 2020). States have also tried to regulate the use of guns and related behaviors. These policies include laws like Child Access Prevention laws (21 states) that require parents to store guns in a specified manner (Webster et al., 2004; Hepburn et al., 2006; Azad et al., 2020). An alternative direction are "stand-your-ground" laws, present in 30 states, that extend the rights of gun owners to use deadly force for self-defense in public spaces, even in cases where retreating from a conflict would be sufficient (Schell et al., 2020; Yakubovich et al., 2021; Degli Esposti et al., 2022). Some states have laws that place limits on the design and manufacture of guns, including regulations of private manufacture. Many states have laws that prohibit guns in certain locations, like school grounds, public transit, or government buildings. Among many other laws, states have also traditionally had laws that regulate who can carry a concealed handgun in public.

Researchers have taken advantage of state-level variation in these laws to study the relationship between their enactment and different measures of gun violence. Most of these studies focus on mortality data, with most homicide or suicide data acquired from sources like the National Vital Statistics System. Other studies use crime data, usually acquired from the FBI's historic Uniform Crime Reports or from the newer National Incident-Based Reporting System. For state laws, researchers either identify and code them through their own legal research or use an existing database (e.g., Siegel et al., 2017). Though many studies are cross-sectional or otherwise descriptive, researchers attempting to make causal inferences have used strategies involving panel data, with policies changing over time. The chief goal of this chapter is to understand how best to evaluate policies with this kind of data.

Studies using panel data have generated a body of scholarship that reveals some trends in the relationship between gun violence and gun policy. For example, multiple studies of laws requiring a license or permit to purchase a firearm have found reductions in firearm homicide, firearm suicide, mass shooting incidence, and other measures of gun violence associated with enactment of these laws (McCourt et al., 2020; Crifasi et al., 2015, 2018; Hasegawa et al., 2019; Webster et al., 2014, 2020). Similar studies on laws requiring a background check at the point of sale (without any license requirement) have failed to find any consistent effect on population-level violence (McCourt et al., 2020; Castillo-Carniglia et al., 2018, 2019; Kagawa et al., 2018, 2023). Child Access Prevention Laws have consistently been associated with reductions in suicide, unintentional shootings, and interpersonal violence. Studies of stand-your-ground laws, on the other hand, have regularly found associations with increased violent crime (Smart et al., 2023).

7.1.2 Plan for the Chapter

In this chapter, we will go through the important elements of impact evaluations of firearm policies using modern methods. First, in Section 7.1.3, we will describe a case study on firearm policy evaluation that will serve as a running example throughout the chapter. We then separate out our development into three key steps. First and foremost, in Section 7.2 we discuss the important design considerations that allow us to be clear about *what* effects we are attempting to estimate. Second, in Section 7.3 we will begin our discussion of *how* to estimate these effects by focusing on estimating effects for a single unit in comparative case studies. Third, we turn to estimating and aggregating effects for multiple units in Section 7.4. In Section 7.5 we discuss additional methodological challenges for firearm policy impact evaluations. Throughout this chapter we pair a formalization of the statistical issues with a practical discussion about implementation.

7.1.3 Case Study: The Impact of Right-to-Carry Laws on Violent Crime

To guide us through the key methodological considerations for estimating the impact of firearm policies, we will focus on a particular case study: measuring the impact of so-called "right-to-carry" (RTC) laws. Many gun policies, including those described above, have been the subject of multiple studies, but perhaps no policy has been examined for as long – and with such politicization – as RTC laws. Academic researchers have been evaluating these laws, otherwise referred to as "shall-issue" or "permitless" concealed carry weapons (CCW) laws, since the mid-1990s. As we will discuss below, the policy space has changed significantly over the last 30 years; many states have enacted a form of RTC law, giving further policy variation over states and time that we can leverage to estimate policy impacts.

RTC laws generally intend to increase the number of citizens who have the right to carry a concealed weapon in public (we defer further policy details to Section 7.2). The changing landscape of state laws governing concealed gun carrying has provided researchers an opportunity to estimate the relationship between changes to these laws and public safety. Early research on CCW laws by Lott and Mustard (1997) suggested that RTC law adoption led to reductions in violent crime. The "more guns, less crime" hypothesis that was foundational to this work posited that if more civilians were legally carrying firearms in public, criminals would be deterred from engaging in crime as they would not know if a potential target was armed. The study, however, contained serious methodological flaws noted in the literature (Webster et al., 1997; Doucette et al., 2019; Donohue et al., 2019). Despite these flaws, this hypothesis and perceived deterrent benefits have been used to argue for lowering restrictions for concealed carry.

More recent research, with access to additional years of data and using newer statistical techniques, has evaluated the impacts of adopting RTC laws on violent crime outcomes (e.g., McElroy and Wang, 2017; Siegel et al., 2017, 2019; Crifasi et al., 2018; Knopov et al., 2019; Doucette et al., 2019, 2022, 2023; Van Der Wal, 2022; Donohue et al., 2023; Smart et al., 2023). Broadly, these studies have found that RTC laws – laws that make it easier to carry a concealed gun in public – are associated with different measures of violence, including violent crime overall, violent crime in urban counties, officer-involved shootings, and others. Donohue et al. (2023) is one of the only studies to look at the mechanism for these increases, with findings suggesting a link between RTC laws and increases in gun theft and decreases in violent crime clearance rates. In this chapter, we will re-analyze the impact of RTC laws on violent crime (following the data setup in Donohue et al. (2019) in particular) in order to pull out the important steps for gun policy impact evaluations.

7.2 Operationalizing an Impact Evaluation

To begin our discussion, we will go over key considerations for operationalizing an observational impact evaluation with panel data. These decisions are independent of the particular estimation strategies that we discuss in the succeeding sections and should be made before considering estimation.

7.2.1 Defining Units and Exposures

First, we must define the units in our study and the policy levels they are exposed to. An initial step is to consider the jurisdictional level where the policy change of interest occurs. For example, RTC laws are state-level policy changes, and so it is appropriate to set our units of analysis to be states in the US. Therefore, throughout this chapter we will use "units" and "states" interchangeably to align with the running example of evaluating RTC laws, and we will say that we have N states, indexed by $i = 1, \ldots, N$. However, our discussion is general to any unit of analysis.[1] We also assume that the data is available at the state level, or that it has already been aggregated to the state level.

Next, we must use a consistent definition of the policy levels, so that it is clear what we mean when we say that two states have the same "treatment." In particular, in order for any policy impacts to be interpretable as causal effects, there cannot be multiple versions of policies – or if there are, they should be characterized as different "treatments." This, along with the no interference assumption discussed below, is often referred to as the Stable Unit Treatment Value Assumption (SUTVA) (Rubin, 1980). Identifying what different versions of the policy of interest exist often requires legal research and qualitative methods to categorize policies; sometimes those who have conducted such research make their categorizations available as shared databases.

Except in special cases, even if two states have the "same" policy, the particulars of the implementation may differ greatly. This creates a practical challenge for policy evaluations. For example, returning to right-to-carry laws, historically there have been four different types of concealed carry weapons law: "no-issue," "may-issue," "shall-issue," and "permitless" carry laws (Doucette et al., 2023; Donohue et al., 2023). Under no-issue CCW laws, concealed carry is effectively banned. May-issue laws allow individuals to apply for a CCW permit, but give law enforcement broad discretion over who is eligible to receive one, making obtaining a CCW permit difficult. In contrast, shall-issue laws remove law enforcement discretion, and permitless CCW laws remove any form of vetting or licensing to carry concealed firearms in public. These latter two policies are often combined under the "right-to-carry" title because both make it significantly easier to carry a concealed gun in public spaces. This leads to a trade-off between "lumping" policies that are broadly similar together and "splitting" policies apart based on their particulars. The splitting approach allows for a more coherent definition of the treatment levels; however, this reduces the amount of data we have per treatment level and increases the number of potential comparisons combinatorially. As a result, researchers often "lump" policies together (Hasegawa et al., 2020).

We follow the original analysis of Donohue et al. (2019) and "lump" shall-issue and permitless systems together into a broad RTC treatment, consistent with the typical framing of these laws. As a result, any estimated effects can be thought of more properly as an average over the effects of shall-issue and permitless policies, a less interpretable measure but one with more statistical power to measure effects. With this definition, in 2024 the vast majority of states have enacted some form of RTC law, while in 1980 the vast majority of states had not.[2]

7.2.2 Defining Outcomes of Interest

Having defined the units (states) and the exposure levels (a broad notion of RTC), we move on to defining our outcomes of interest and their timing. In general, most previous firearm policy evaluations have used either fatal firearm violence from the CDC's National Vital Statistics System (NVSS) or violent crime data from the FBI's Uniform Crime Report (UCR). The CDC's NVSS provides counts of homicides, suicides, and unintentional deaths by mechanism (i.e., firearm or non-firearm) for all 50 states going back several decades using death certificate data. While this data is publicly available from the CDC's WISQARS platform, that data is subject to censorship when the counts of a given state-year are below 10. This limits the usability of the publicly available dataset and necessitates that researchers apply for, obtain, and clean raw data provided by the NVSS. In order to obtain the raw NVSS data, one must (i) hold a terminal degree (e.g., PhD, ScD, MD, etc.); (ii) articulate a clear research strategy for data use; and (iii) agree to strict data storage and destruction requirements. However, because the raw NVSS data rely on death certificates, there is a relatively low level of bias due to under-counting.

In contrast, the FBI's UCR is an uncensored, publicly available data set. However, it relies on self-reported data and likely contains moderate to high levels of bias. The UCR contains data reported by over 15,000+ local police jurisdictions and provides data on offenses known to law enforcement (e.g., crimes reported to law enforcement), arrests made, and other criminal activities, such as the number of hate crimes or arsons (Banks et al., 2016). When using UCR data for policy evaluation, it is important to consider the issue of missing data. In particular, larger jurisdictions are more likely to report data: the UCR relies on local law enforcement jurisdictions to self-report data, and larger jurisdictions have greater statistical capabilities to do so. Moreover, the FBI contacts jurisdictions representing populations of 100,000 or more to increase data reporting (Maltz, 2006). When using UCR data for state-level policy evaluation, it is important to give thought to whether and how the proportion of local jurisdictions that report data in a given state and year changes over time. A large fluctuation in reporting across states and years may be a result of large changes in the number of crimes or an artifact of changes in reporting practices.

In addition, recent changes to crime data reporting may complicate future analyses. In 2021, the FBI began requiring that law enforcement agencies report crime to the National Incident-Based Reporting System (NIBRS) instead of the UCR. NIBRS requires more incident-level data than UCR which means it is ostensibly a richer data source (Lantz, 2022). Unfortunately, implementation of the 2021 switch has been messy. In 2021, for example almost 40% of the nation's 18,000 law enforcement agencies failed to report to NIBRS (Li, 2022). The proportion improved in 2022, with 31% of agencies failing to report data (Li et al., 2023), and FBI data suggests reporting coverage continues to improve. Despite these improvements, the reporting gaps during the transition to NIBRS have triggered concerns about the reliability of crime data from 2021 (Asher, 2022; Parker, 2022). Researchers seeking to use FBI crime data will have to carefully analyze the completeness of the data for jurisdictions included in the study and account for the possibility of incomplete data.

With these caveats in mind, for our running example we will follow Donohue et al. (2019) and evaluate the impact of RTC on violent crime (defined as murder, rape, robbery, and aggravated assault), using the FBI UCR data and dividing the number of violent crimes by state's population size. We note that because of the potential issues with the UCR data, one should view this as a stylized analysis that elucidates the key methodological ideas.

One final consideration is the frequency with which outcomes are measured. We will focus on yearly crime measurements in our running example, so that we observe 38 crime rate measurements for each state from the years 1977 to 2014. Finally, as a bit of notation, we will denote the violent crime rate per 100,000 residents for state i in year t as Y_{it}.

7.2.3 Defining Time Zero

Next we must define a notion of "time zero" for each state: the year that a state may or may not have enacted an RTC law. The key is to determine what are pre-treatment baseline measures and what are post-treatment outcome measures. This is easy for the states that did actually implement the policy but more challenging for the states that did not. However, it is a crucial step as inadvertently conditioning on post-treatment values can lead to bias, even more than failing to adjust for potential confounders.

In public policy contexts, states can typically implement a policy change at any time.[3] For a treated state that eventually enacts an RTC law, we can define "time zero" as the year that the policy was enacted. Formally, for a state j, we use T_j to denote that year RTC was enacted, and we measure outcomes for a given "event time" k centered around this year, Y_{jT_j+k}. Negative event times $k<0$ denote pre-treatment measures, $k=0$ denotes the year that RTC was enacted, and positive event times $k>0$ denote post-treatment measures. For example, Maine enacted RTC in $T_j=1985$, so event time $k=-5$ corresponds to $T_j+k=1980$, event time $k=0$ corresponds to $T_j+k=1985$, and event time $k=5$ corresponds to $T_j+k=1990$.

Figure 7.1 shows which states enacted RTC in what year. There are several salient features. First, we see in Figure 7.1 the tell-tale "staircase" showing staggered adoption of treatment: no state that enacted RTC repealed it during our panel from 1977 to 2014. Second, we see that five states (NH, WA, CT, VT, and AL) had already enacted RTC before our first year of consistent data in 1977, and so pre-treatment measures with negative event times $k<0$ do not exist. As we will be using methods that require pre-treatment measures, we will have to remove these five states from our sample. In addition, we remove Indiana (RTC 1980) from our sample because there are only three years of pre-treatment data (i.e., $k=-3,-2,-1$), which is insufficient for

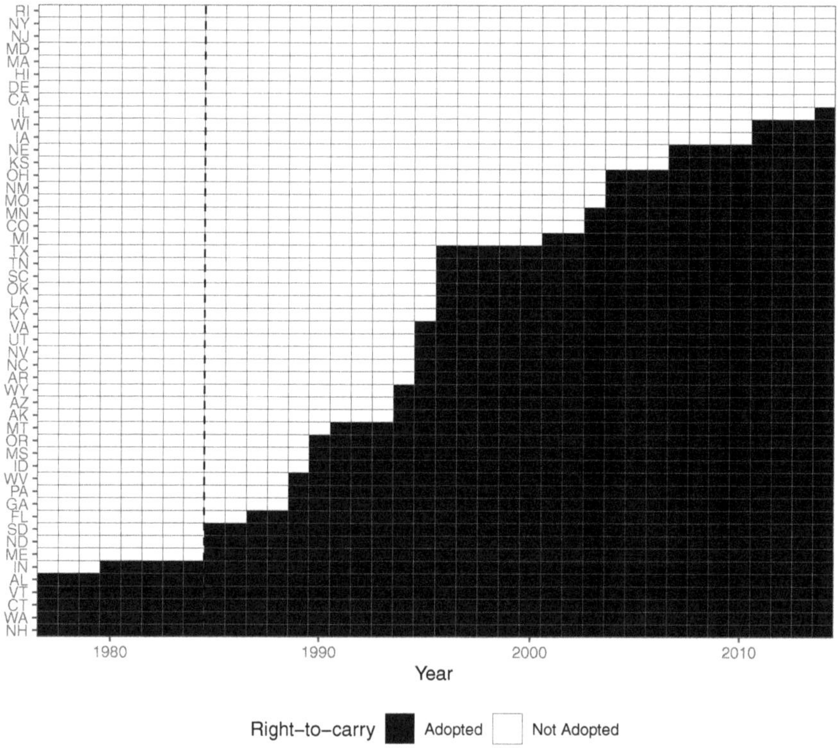

FIGURE 7.1
Timing of Right-to-Carry Laws Across the United States.

identifying the pre-treatment trends that we will need to adjust for in later sections in order to reliably estimate effects. Finally, eight states (CA, DE, HI, MD, MA, NJ, NY, and RI) did not enact RTC during the period 1977–2014. These are so called "never treated" states, and we denote their treatment time as infinity, $T_j = \infty$. These never-treated states have no obviously defined time zero, but they will serve as comparators to states that enacted RTC during 1977–2014 in our analysis.

7.2.4 Defining Causal Contrasts of Interest

With the various aspects of our dataset in place, we can now turn to defining our causal contrasts of interest. To do so, we will use the potential outcomes framework (Neyman, 1923; Rubin, 1974), as adapted to this data setting by Athey and Imbens (2022). First, we need to create an expanded notion of the "treatment." Recall that we have "lumped" the intervention into a broad RTC category so that there are two conditions for a state: an RTC law has been enacted, or it has not. However, the timing of the policy change is also potentially important, and so we must track whether and when a state is treated in our analysis, using the treatment time variable T_i defined above.

We will consider the potential violent crime rate for state i in year t if it were to enact RTC in year s, denoted as $Y_{it}(s)$. In order to link these potential crime rates to the crime rates that we actually observe, we rely on the following two assumptions.

1. **No interference between states:** a state is only affected by its own RTC status. If there are spillovers across states and this assumption is violated, it becomes difficult to even define notions of causal effects. In practice while we might expect spillovers to exist, the no interference assumption may be a reasonable approximation.
2. **No anticipation:** prior to enacting RTC, a state's potential violent crime rate is equal to its rate if it were never treated. We formalize this assumption as $Y_{it}(s) = Y_{it}(\infty)$ for $t < s$, with treatment time s. This assumption is sometimes highly implausible, for instance, if the announcement of a policy changes behavior before the policy itself takes effect. If such anticipation is a concern, we can re-define the treatment time to be an earlier date.[4]

Under these two assumptions we can link the observed outcomes to the potential outcomes as follows: if a state i enacts RTC in year T_i, the outcomes for years prior to T_i correspond to the "never-treated" potential outcome $Y_{it} = Y_{it}(\infty)$ and the outcomes for years T_i and later correspond to the potential outcome when enacting an RTC law in year T_i, $Y_{it} = Y_{it}(T_i)$.

With all of this formalization, we can finally discuss what we mean by a "causal effect." There are many potential comparisons and counterfactuals. For example, we could focus on a single state, like Ohio (RTC 2004), and compare its violent crime rate in 2000 in a world where it had previously implemented RTC (say, in 1995) to a world where it had not yet implemented RTC. These types of comparisons quickly become unwieldy. A simple comparison often used in practice is to compare the observed crime rate in an RTC state to its counterfactual crime rate in a world where it never enacts RTC (formally, $Y_{it}(T_i) - Y_{it}(\infty)$). To facilitate comparisons across states, we will turn to our discussion in Section 7.2.3 and measure this effect of RTC at event time k years after adopting RTC. Formally, we write this effect as $\tau_{ik} = Y_{iT_i+k}(T_i) - Y_{iT_i+k}(\infty)$. Because we observe what happens under RTC ($Y_{iT_i+k}(T_i)$), the central methodological challenge is to impute what would have happened had the state not yet enacted RTC ($Y_{iT_i+k}(\infty)$).

While we are sometimes interested in estimating this state-level causal effect, it is difficult and noisy to estimate in practice. Instead, we will focus on estimating an aggregate effect, the (sample) Average Treatment Effect on the Treated states (ATT), k years after each state adopted RTC:[5]

$$ATT_k = \frac{1}{N_{trt}} \sum_{i|T_i<\infty} \tau_{ik},$$

where N_{trt} is the number of states that ever adopted RTC. This aggregate effect aligns the states according to event time, so that, e.g., ATT_0 corresponds to the average effect of RTC across states that enacted it, in the year that it was adopted, relative to a counterfactual scenario where the state had not adopted RTC by that year.

7.3 Building Blocks: Estimating Effects in Comparative Case Studies

Only now that we have carefully considered *what* policy effects we want to estimate, and how to interpret them, can we turn to *how* to estimate them. In this section we will introduce the basic building blocks of two main approaches to estimating policy effects – differences in differences and synthetic controls – and will focus on using each estimator to measure the impact of RTC in a single state, which we can view as a comparative case study. Although these state-level estimates are noisy, they develop the key concepts for both approaches, which we then show how to aggregate into a full analysis in Section 7.4. The two states we use as running examples are Ohio and West Virginia; each has aspects of their analysis that highlight important points for estimation.

7.3.1 Selecting Units for Comparison

The first key choice in the analysis is carefully selecting comparison states for the focal state of interest. Our first concern is ensuring that we are making valid comparisons, comparing the focal state to other states that either never enact RTC or enact RTC after the period of interest. States that have already adopted RTC by the period of interest cannot serve as comparisons without strong (and typically unreasonable) modeling assumptions. With such a comparison set of "donor states," we can try to statistically adjust for any differences between the donors and the focal state.

An important consideration in selecting comparison states is the length of the period for which we want to measure effects. Formally, we are interested in estimating effects for K years after treatment. We have two options for constructing the donor set. First, for each event time $k = 0, 1, \ldots, K$, we could construct a separate donor set consisting of all states that have not yet enacted RTC. This gives us the largest possible number of donors, but differences in impact estimates each year may be due to differences in the composition of the donor set, which would have to be carefully checked. In this chapter, we take an alternative approach: keeping the donor set the same for each treatment year by only including those states that do not enact RTC in the K years after the focal state.

Formally, we define the set of possible donor units for treated state j as those states i which are treated at least $K+1$ periods after state j, which we denote as $D_j \equiv \{i : T_i \geq T_j + K + 1\}$. We will use N_j to denote the number of donor units in D_j. With this setup, the longer the post-period length K, the fewer donor states are available. If we choose a long enough post-period length, only never-treated states will serve as valid comparators, while setting a shorter post-period length allows states that are late adopters to serve as comparisons for early adopters.

In this chapter, we focus on effects up to $K = 10$ years after a state enacts a right-to-carry law. For example, West Virginia enacted RTC in 1989, and we measure effects until 1999; for Ohio (RTC in 2004) we measure effects until 2014. For states that do not have $K = 10$ years of post-RTC data, we include as long of a post-period as possible. For example, for Wisconsin (RTC 2011) we estimate effects for 3 years until 2014, and for Illinois, which enacted RTC in the final year of the panel (2014), we measure effects only for 2014.

7.3.2 Difference-in-Differences

The workhorse method for estimating policy effects with panel data is the Difference-in-Differences method (DiD). The key idea of this approach is to compare the difference between the focal and comparison states before and after the policy change. If there is a change in the difference after the policy is enacted, this could be evidence of an effect. See Angrist and Pischke (2008) §5 for a textbook treatment on difference-in-difference methods and Roth et al. (2023) for an in-depth review. To build intuition for the DiD estimator, we first consider two problematic comparisons and discuss how DiD addresses the limitations of each. Figure 7.2a shows the violent crime rate for Ohio and donor states from 25 years before and 10 years after Ohio enacted a right-to-carry law (2004), along with the average for the donor states (Table 7.1).

The first comparison is between the violent crime rates in Ohio and its comparison states in the ten years following Ohio enacting its RTC law in 2004. During these ten years, Ohio saw an average of 326 crimes per 100,000 residents, 21% lower than the comparison states, which had an average of 417 crimes per 100,000. Drawing causal conclusions from this simple comparison would ascribe all of the difference between Ohio and its comparison states to the RTC law, but Figure 7.2a shows that in fact Ohio had a lower violent crime rate than its comparison states even before enacting its RTC law.

As an alternative, we can instead inspect the violent crime rate in Ohio before and after enacting RTC. In the 25 years preceding the policy change, Ohio had an average of 432 violent crimes per 100,000 residents. In the 10

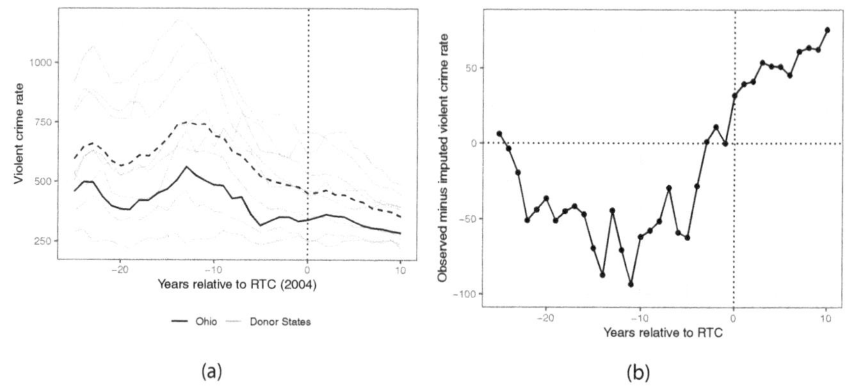

FIGURE 7.2
(a) The Violent Crime Rate for Ohio (Black) and Donor States (Gray) from 25 Years Before to 10 Years After Ohio Enacted a Right-to-Carry Law. The dashed line represents the average violent crime rate among the donor states. (b) An "Event Study Plot" of the Difference-in-Differences Estimate of the Impact of RTC on the Violent Crime Rate in Ohio Pre- and Post-RTC.

TABLE 7.1
A Two-by-Two Difference-in-Differences Table for Ohio

	Right-to-Carry Law		
	Pre (1979–2003)	**Post (2004–2014)**	**Difference**
Donors	616.95	417.25	–199.70
OH	431.88	326.19	–105.69
Difference	–185.07	–91.06	94.01

Note: The average violent crime rate per 100,000 residents in Ohio and its comparison states in the period before and after Ohio enacted RTC.

years afterwards there were an average of 326 violent crimes per 100,000 residents, a reduction of 24%. However, drawing causal conclusions from this comparison requires us to believe that if Ohio did not enact an RTC law, the violent crime rate would have remained unchanged from its historical average. Looking again at Figure 7.2a, this seems implausible given that there was a reduction in the violent crime rate for all the donor states during the post-period as well.

The DiD estimator combines these two approaches by comparing the average difference between Ohio and its comparison states *after* enacting RTC (–91 violent crimes per 100,000) to the difference *before* enacting RTC (–185) to estimate an increase of 94 violent crimes per 100,000 in the post-RTC period (see Table 7.1). We can view this as using the pre-RTC difference to adjust for pre-existing dissimilarities between Ohio and its donor states. An equivalent way to find this estimate is to compare the change in the average violent crime rate in Ohio before and after it enacted its RTC law in 2004 (–105.7), to the change in the violent crime rate in the donor states during the same period (–199.7). We can view this characterization as using the pre-post difference in the donor states to adjust for any common changes in the violent crime rate across all states.

After taking the difference of the differences, we are left with a treatment effect of 94 violent crimes per 100,000 per year that may be ascribed to RTC. The key assumption underlying the validity of the DiD estimator is a *parallel counterfactual trends* assumption. This assumes that in the absence of Ohio enacting an RTC law in 2004, the trend in the violent crime rate would match the trend over that same time period in its comparison states (Card and Krueger, 2000; Angrist and Pischke, 2008). Note that this assumption depends on our choice of scaling of the outcomes. For example, if there are parallel trends for the violent crime rate, then there typically will not be parallel trends for the log violent crime rate or any other non-linear transformations without several stronger assumptions.

A threat to the parallel trends assumption is the existence of time-varying confounding factors. By comparing to the pre-RTC differences between the states, the DiD estimator can adjust for any fixed dissimilarities, even if they are not observed. However, if the states differ in ways that are not stable over time such as relative changes in demographic profiles or if the policy implementation process depends on features other than baseline violent crime rates – e.g., changing policy in response to a particular spike in the crime rate in a particular year – the parallel trends assumption will not hold.

The two-by-two comparison outlined here attempts to measure the impact of RTC averaged over the entire post-RTC period. However, we can adapt it to estimate effects for each individual post-RTC year. To do so, we fix a common reference year, typically the year immediately preceding the policy change ($T_j - 1 = 2003$, in this case), then for every other year we compute the DiD estimator as above, using 2003 as the "pre-period" and using each other year as the "post-period." Formally, we can write this estimator as

$$\hat{\tau}_{jk}^{did} = Y_{jT_j+k} - Y_{jT_j-1} - \frac{1}{N_j}\sum_{i \in \mathfrak{D}_j}\left(Y_{iT_j+k} - Y_{iT_j-1}\right). \tag{7.1}$$

Figure 7.2b shows these estimates in what is often called an "event study" plot. In addition, we can formalize the parallel counterfactual trends assumption as assuming that the expected change in the violent crime rates between the last pre-RTC period and every post-RTC period for treated state j is equal to that for the donor states:[6]

$$\mathbb{E}\left[Y_{jT_j+k}(\infty) - Y_{jT_j-1}(\infty)\right] = \mathbb{E}\left[\frac{1}{N_j}\sum_{i \in \mathcal{D}_j} Y_{iT_j+k}(\infty) - Y_{iT_j-1}(\infty)\right] \quad \textit{for all } k \geq 0. \tag{7.2}$$

This more flexible estimation strategy allows us to construct a diagnostic measure for the parallel counterfactual trends assumption. The estimates $\hat{\tau}_{jk}^{did}$ for $k < 0$ measure the impact of RTC *before* it is enacted. These serve as "placebo" estimates because they should be near zero; estimating non-zero impacts before the policy change occurs is a warning sign that at least one key assumption is violated. Figure 7.2b shows that there do not appear to be parallel trends, instead the event study plot is indicative of "pre-trends": about ten years before enacting RTC (i.e., in 1994) the violent crime rate in Ohio began to reduce more slowly than in its comparison states, so that we see a stark upward trend in the DiD estimates. This indicates that the parallel trends assumption or another key assumption such as the lack of anticipation is likely to be violated, and so we should not trust these estimates.

Note that this diagnostic is not a direct test of the parallel counterfactual trends assumption because the assumption is a statement about unobservable *post-intervention* counterfactual trends. However, it is a useful proxy: if there is evidence that trends are not parallel in the pre-RTC period, then it is more difficult to argue that the counterfactual trends in the post period are parallel.[7] Furthermore, as with any statistical test, lack of evidence against the parallel trends assumption is not evidence for parallel trends (see Hartman and Hidalgo, 2018, for further discussion on equivalence tests for placebo checks). One should not use noisy estimates around zero as a substitute for arguing why the assumptions hold.

7.3.3 When Parallel Trends Fail: Synthetic Controls

The event study plot for Ohio in Figure 7.2b precludes us from taking the DiD estimates seriously because there is evidence of differential trends in the pre-RTC period – the violent crime rate in Ohio appears to have been on a different trajectory than the average of the other states. As discussed, the DiD approach begins with a flawed comparison of the difference in the violent crime rate between Ohio and the average of the comparison units and then attempts to correct for pre-RTC differences in the violent crime rate with the second difference. An alternative approach is to alter the set of comparison states in some way so that the average violent crime rate for the altered set of comparison states is a good proxy for the violent crime rate in Ohio. The synthetic control method (SCM), proposed by Abadie and Gardeazabal (2003) and further developed in Abadie et al. (2010, 2015), is an optimization-based procedure to alter the comparison set in a data-driven way.

As a first step, we must choose what it means to "alter" the comparison set. One possibility is *matching*, which entirely removes some states from the comparison set. We will instead consider differentially *weighting* states in the comparison set so that we compare Ohio to a weighted average of states, $\sum_{i \in D_{OH}} \gamma_{iOH} Y_{it}$, where γ_{iOH} is the weight placed on comparison state i. This weighted average of the comparison states is often referred to as a *synthetic control* state; by up-weighting some states and down-weighting others, we can try to find a synthetic control that will be a good proxy for what would have happened in Ohio absent the enactment of RTC.

To find such a synthetic control, recall that the issue with comparing Ohio to the raw average of the comparison states is that post-RTC differences may simply reflect pre-RTC differences, as we see in Figure 7.2a. The key idea of SCM is to target this issue directly and find weights so that the synthetic control state has a violent crime trajectory that resembles that in Ohio during the pre-RTC time period. For a pre-treatment time $t < 2004$, we can evaluate the synthetic control by looking at the difference between the observed violent crime rate in Ohio and the value imputed by the synthetic control: $Y_{OHt} - \sum_{i \in D_{OH}} \gamma_{iOH} Y_{it}$. This gives us a snapshot of the pre-RTC difference between Ohio and the synthetic control. To get the full picture, we aggregate across the pre-RTC period by taking the mean square difference, often called the Mean Square Prediction Error (MSPE):

$$\frac{1}{24} \sum_{t=1979}^{2003} \left(Y_{OHt} - \sum_{i \in \mathcal{D}_{OH}} \gamma_{iOH} Y_{it} \right)^2$$

where $2003 - 1979 = 24$ is the number of pre-RTC years for Ohio in our data. We can then try to find the optimal synthetic control that minimizes this average squared difference (Figure 7.3).

We can take this approach for any treated state j, by trying to find weights $\hat{\gamma}_j = (\hat{\gamma}_{1j}, \ldots, \hat{\gamma}_{Nj})$ that minimize the MSPE during the pre-RTC period, the length of which we denote by L_j. Formally, the SCM weights $\hat{\gamma}_j$ solve the following constrained optimization problem:

$$\min_{\gamma_{1j}, \ldots, \gamma_{Nj}} \frac{1}{L_j} \sum_{\ell=1}^{L_j} \left(Y_{jT_j - \ell} - \sum_{i \in \mathcal{D}_j} \gamma_{ij} Y_{iT_j - \ell} \right)^2 \tag{7.3}$$

subject to $\sum_i \gamma_{ij} = 1,\ \gamma_{ij} \geq 0,\ \gamma_{ij} = 0 \textit{ for } i \notin D_j$.

The objective function of this optimization problem corresponds to the MSPE for a synthetic control defined by weights $\gamma_{1j}, \ldots, \gamma_{Nj}$ during the pre-RTC period for treated state j. In addition to this objective, we include constraints on the weights that ensure that the weights (i) are non-negative ($\gamma_{ij} \geq 0$ for all i); (ii) sum to one ($\sum_i \gamma_{ij} = 1$); and (iii) exclude impossible donors ($\gamma_{ij} = 0$ whenever i is not a possible donor, $i \notin D_j$). With these weights, the SCM estimate of the treatment effect compares the observed violent crime rate to the violent crime rate imputed by the synthetic control:

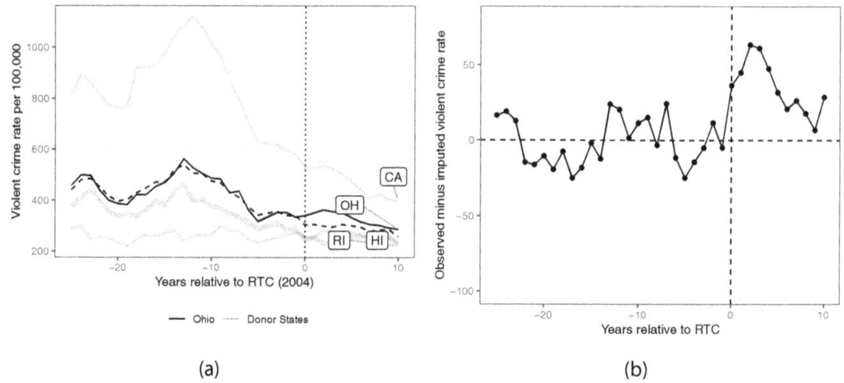

FIGURE 7.3
(a) The Violent Crime Rate for Ohio (Black Solid Line) and Donor States that Form the Synthetic Control (Gray, Size Proportional to Weight), and the Imputed Violent Crime Rate for Its Synthetic Control (Black Dashed Line) from 25 Years Before to 10 Years After Ohio Enacted a Right-to-Carry Law. (b) A "Gap Plot" Showing the SCM Estimates of the Impact of RTC on the Violent Crime Rate in Ohio Pre- and Post-RTC.

$$\hat{\tau}_{jk}^{scm} = Y_{jT_j+k} - \sum_{i \in D_j} \hat{\gamma}_{ij} Y_{iT_j+k}. \tag{7.4}$$

A key feature of the SCM optimization problem (1.3) is the constraint that the weights are non-negative and sum to one. This constraint stops the synthetic control from *extrapolating* away from the support of the donor units. It has the additional effect that typically most of the SCM weights will be zero, and so the synthetic control will be a composite of only a few donor states, allowing for an interpretable comparison.

Returning to our running example in Ohio, Figure 7.3a shows the violent crime rate per 100,000 residents in Ohio and its donor states, as in Figure 7.2a, now with the donors' violent crime rate series sized proportional to their weight in Ohio's synthetic control. The main states that make up the synthetic control are Rhode Island (60%), Hawaii (20%), and California (20%); the other states contribute less than 1% to the synthetic control. Taking the weighted average of these states gives the imputed violent crime rate from the synthetic control.

By finding an optimal combination of donor states, the synthetic control method allows us to weaken the assumption of parallel counterfactual trends on which the DiD estimator relies. SCM allows for the presence of unmeasured time-varying confounders; however, it requires that the relationship in the violent crime rate between states is stable over time.[8] In this setting, SCM will provide good estimates of the treatment effect under two conditions: (i) the number of pre-treatment periods should be large, and (ii) the pre-treatment fit, i.e. the difference between the violent crime rate in Ohio and its synthetic control in the pre-RTC period must be small in magnitude (Abadie et al., 2010). Both conditions are important to ensure that the synthetic control is not biased: if the pre-period is short, then the synthetic control may be over-fitting to noise; if the pre-treatment fit is poor, then the synthetic control may be under-fitting to the signal.

While heuristics and rules of thumb for these two conditions are not readily available, to evaluate the quality of the synthetic control we can use the same strategy as with the DiD estimator above: the SCM estimates of the impact of RTC in the pre-RTC period ($\hat{\tau}_{jk}^{\text{scm}}$ with $k < 0$) should be near zero, and so they can again serve as placebo estimates. We can visualize this via an analog to the event-study plot in Figure 7.2b: plotting $\hat{\tau}_{jk}^{scm}$ against the number of years before ($k < 0$) and after ($k \geq 0$) the implementation of RTC in Ohio. In synthetic control settings, this is often called a "gap plot," shown in Figure 7.3b.[9] We can see that the violent crime rate imputed by the synthetic control closely follows the observed violent crime rate in Ohio, but the fit is not perfect. A useful one-number summary to diagnose the fit is the root mean square prediction error (RMSPE), the square root of the MSPE, which is equivalent to the square root of the objective in the SCM optimization problem (1.3). Using this metric, we see that the violent crime rate in Ohio differs from its synthetic control by roughly 17 violent crimes per 100,000 in the pre-period.

7.3.4 When Pre-treatment Fit Is Poor: Bias Correction

While the pre-treatment fit for Ohio is reasonable, it will not always be the case that it is possible to find a good synthetic control for a state. In fact, when studying the impact of a policy enacted in multiple states, it is likely that there is no good synthetic control for at least some of them. As an example of this, consider West Virginia (RTC in 1989). Following the procedure above gives a synthetic control for West Virginia comprised of Wisconsin (58%) and Iowa (42%), but this synthetic control is a poor fit. Figure 7.4b shows the gap plot. The RMSPE is 52 violent crimes per 100,000, and the gap plot shows that in the years leading up to enacting RTC in 1989, WV began to differ substantially from its synthetic control. In 1988, immediately prior to enacting RTC, WV had 100 fewer violent crimes per 100,000 than its synthetic control. These estimates are not credible, but by construction we have found the best possible synthetic control. We now turn to what to do in these settings.

The lack of pre-treatment fit for West Virginia exhibited in Figure 7.4b indicates that the synthetic control estimator is likely to be biased. To get a sense of how big this bias can be, we can attempt to directly estimate it. We can do this by first fitting a predictive model of the violent crime rate in the post-RTC period, if RTC were not enacted, using the donor states for WV. Then we can evaluate the difference between the predicted crime rate in WV and the weighted average of the predicted crime rates in the component states

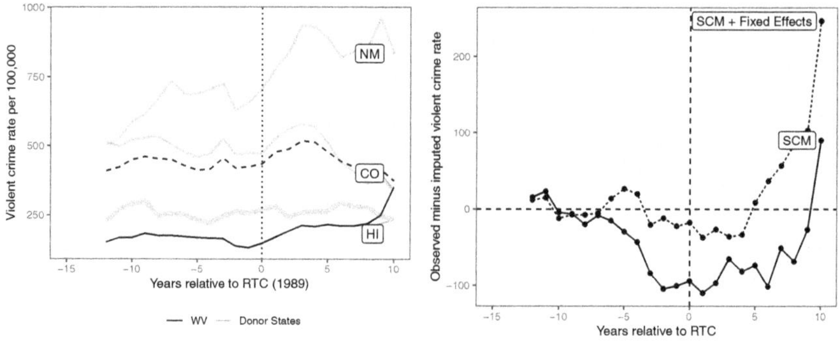

(a) (b)

FIGURE 7.4
(a) The Violent Crime Rate for West Virginia (Black Solid Line) and Donor States that Form the Synthetic Control (Gray, Size Proportional to Weight), and the Imputed Violent Crime Rate for Its Synthetic Control (Black Dashed Line) from 12 Years Before and 10 Years After WV Enacted a Right-to-Carry Law. (b) A "Gap Plot" Showing the SCM and Fixed-Effects Augmented SCM Estimates of the Impact of RTC on the Violent Crime Rate in WV Pre- and Post-RTC.

of its synthetic control (WI and IA). The difference between these two is the model's prediction about the bias of the SCM estimate. Formally, if $\hat{Y}_{it}(\infty)$ is a prediction of the violent crime rate in the absence of RTC, then the estimate of the bias is

$$\widehat{bias}_{jk} = \hat{Y}_{jT_j+k}(\infty) - \sum_{i \in D_j} \hat{\gamma}_{ij} \hat{Y}_{iT_j+k}(\infty).$$

While this bias estimate can serve as a diagnostic tool, we can also use it to directly adjust the SCM estimate to correct for the bias by simply subtracting out the estimate of the bias from the estimate of the treatment effect (Ben-Michael et al., 2021a). The bias-corrected (or augmented) synthetic control estimate is

$$\hat{\tau}_{jk}^{ascm} = \hat{\tau}_{jk}^{scm} - \widehat{bias}_{jk}.$$

There are many potential predictive models to use to estimate and correct for the bias. Ben-Michael et al. (2021a) primarily advocate for using ridge regression to fit a predictive model of post-RTC crime rates on pre-RTC crime rates for states that did not enact RTC and then use this to predict the post-RTC crime rate in the absence of RTC for the treated states. For simplicity, in this chapter, we will focus on another predictive model: two-way fixed effects estimation. This procedure estimates a state-level fixed effect $\hat{\phi}_i$ and a year-level fixed effect $\hat{\nu}_t$ to predict the crime rate for state i in year t in the absence of RTC as $\hat{Y}_{it}(\infty) = \hat{\tau}_i + \hat{\nu}_t$. Two-way fixed effects models are intimately related to the DiD approach in Section 7.3.2, and in fact, the fixed-effects augmented synthetic control estimate can be written as a form of a weighted DiD estimate:[10]

$$\hat{\tau}_{jk}^{fe-scm} = \frac{1}{L_j} \sum_{\ell=1}^{L_j} \left[\left(Y_{jT_j+k} - Y_{jT_j-\ell} \right) - \sum_{i \in D_j} \hat{\gamma}_{ij} \left(Y_{iT_j+k} - Y_{iT_j-\ell} \right) \right]. \tag{7.5}$$

The fixed-effects augmented SCM has the same form as the DiD estimator (1.1), except that the pre-post differences in the violent crime rate for comparison units are weighted by the SCM weights. Conversely, setting the weights to be $1/N_j$ above recovers the DiD estimate.

This estimator evaluates the bias by looking at the average difference in violent crime rates between WV and its synthetic control in the pre-RTC period. While this is a relatively simple estimate, it affords us several advantages. Because this simple predictive model gives us estimates for all states in all years, we can fit the synthetic control on the residual violent crime rate $Y_{it} - \hat{Y}_{it}(\infty)$; this is equivalent to modifying the SCM optimization problem to include a state-specific intercept (Doudchenko and Imbens, 2016; Ferman and Pinto, 2021; Abadie, 2021). This directly accounts for differences in the level of the violent crime rate between states, allowing the synthetic control to focus on matching the trends in the violent crime rate. This is particularly helpful in contexts such as measuring impacts on crime rates, when the scale of the outcome varies across states, and so it can be difficult to construct an adequate synthetic control that matches both the level and trends together.

Figure 7.4a shows the violent crime rate in West Virginia, the three states that make up its fixed-effects augmented synthetic control (50% Colorado; 35% Hawaii; 14% New Mexico), and the weighted average of the donor states before adjusting for level differences. The weighted average of these states has a much higher level of the violent crime rate than WV in the pre-RTC period (430 per 100,000 compared to 163 per 100,000 on average); however, the trend in the violent crime rate is a relatively good match. By adjusting explicitly for level differences, the fixed-effects augmented synthetic control is free to use states such as New Mexico and Colorado that saw a slight decline in the violent crime rate during the late 1980s – similar to West Virginia, but that had much higher levels. As the gap plot in Figure 7.4b shows, after adjusting for these level differences, the synthetic control has reasonably good fit, with an RMSPE of roughly 16 violent crimes per 100,000, much better than the un-augmented synthetic control, which has an RMSPE of 52 violent crimes per 100,000.

7.3.4.1 Additional Estimation Approaches

A related approach to estimating effects when the synthetic control fit is poor is the "Synthetic Differences in Differences" approach proposed by Arkhangelsky et al. (2021). This procedure estimates effects using a weighted DiD estimate with two types of weights: (i) weighting units by their synthetic control weight as in Equation (7.5) and (ii) weighting the pre-treatment time periods in the adjustment according to how the pre-treatment periods predict the post-treatment periods. Thus, we can view the estimate as a weighted form of the two-by-two DiD table (e.g., Table 7.1), using a pseudo-comparison group via SCM weighting and a pseudo-pre-treatment time via the time weights. As with the augmented approach above, this can also be seen as a form of bias-correction using a model that predicts post-treatment outcomes in the absence of RTC from pre-treatment outcomes.

Finally, a broad class of estimation strategies directly models the outcomes under no treatment via an interactive fixed effects approach (e.g., Xu, 2017) or matrix completion (e.g., Athey et al., 2021) or combines a low-rank matrix approximation with the synthetic controls objective (e.g., Amjad et al., 2018). These approaches attempt to directly estimate and adjust for underlying latent confounders, taking advantage of modern matrix reconstruction techniques. While these methods use different estimation approaches, the principles of the preceding sections are the same, including the design considerations in Section 7.2 and diagnostic checks such as the event study or gap plots.

7.4 Aggregating Effects under Staggered Adoption of Policies

Now that we understand the fundamentals of the DiD and SCM estimation strategies for a single state, we will turn to estimating aggregate effects across the treated states. Because we have laid the groundwork in the previous sections, these final steps will involve relatively simple and transparent aggregations. Some methodological challenges arise due to aggregating estimates across time, but by aggregating across states, we can get less noisy estimates and can construct confidence intervals for the effects via bootstrap re-sampling methods.

7.4.1 From Individual Case Studies to Time Cohorts

As a first step toward estimating aggregated effects, we will consider estimating effects for *treatment time cohorts,* groups of states that enacted an RTC law in the same year. For example, in 2004, the year Ohio enacted an RTC law, Missouri and New Mexico did as well, and Georgia and Pennsylvania enacted RTC laws in 1989 along with West Virginia. To estimate aggregate effects for the whole treatment time cohort with the DiD or (augmented) SCM approaches above, we can simply average the individual-level estimates.

Formally, we estimate the effect for states enacting RTC in year s,k years after enactment as

$$\hat{\tau}_{sk} = \frac{1}{N_s} \sum_{j|T_j=s} \hat{\tau}_{jk},$$

where N_s is the number of states that enacted an RTC law in year s. For DiD this is a straightforward average, while for the SCM estimators this leads to another design choice: we can either find an individual synthetic control for each state or find a single synthetic control for the average violent crime rate in the treatment time cohort. Typically, it is better to average across the time cohort: this adds precision over the noisy-state level measures, and so the resulting synthetic control will typically both be able to achieve better pre-treatment fit and be less liable to over-fit to noise. For example, if we fit separate fixed-effects augmented synthetic controls for the three states in the 1989 cohort, the RMSPEs are 26 (GA), 3.6 (PA), and 16(WV) violent crimes per 100,000. In contrast, if we fit a single synthetic control for the average of the 1989 cohort, the RMSPE for the average is only 1.8 violent crimes per 100,000.

7.4.2 Combining Results Across Treatment Time Cohorts

With these estimates aggregated to the time-cohort level in hand, we can follow the same idea to estimate the overall average treatment effect on the treated. For each year, we estimate the average treatment effect for that treatment-time cohort, then average across treatment time cohorts (weighting by the size of the cohort). We can write this estimator as

$$\widehat{ATT}_k = \frac{1}{N_{trt}} \sum_{s=1}^{T} N_s \hat{\tau}_{sk}.$$

Because we are aggregating across treatment-time cohorts, we must be careful about the differences between calendar time and event time (recall Section 7.2.3). In particular, early adopters of RTC have fewer pre-RTC years from which to fit our estimators and late adopters have fewer post-RTC years from which to measure effects. So, we must restrict the estimate $\hat{\tau}_k$ to aggregate only over states for which we observe outcomes k years post RTC. For example, we can measure effects for all of the 36 different RTC states in the year that they enacted RTC ($k = 0$); however, even for $k = 1$ years post-RTC, we lose Illinois because it enacted an RTC law at the end of our panel. By $k = 10$ years post RTC the panel is reduced to 31 states (see Figure 7.5a).

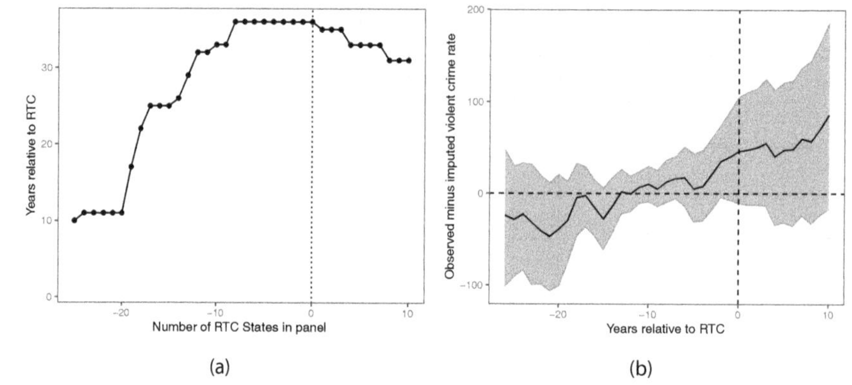

FIGURE 7.5
(a) The number of states that ever enact RTC for which we observe crime rates versus the number of years prior to or after enacting RTC. (b) An aggregated "event study plot" of the difference-in-differences estimate of the impact of RTC on the violent crime rate, along with 95% bootstrap confidence intervals.

7.4.2.1 Stacked DiD

When we take this approach with the DiD estimator, $\widehat{ATT}_k^{\text{did}}$ is a "stacked" difference-in-differences approach that aggregates and combines the individual or time-cohort level two-by-two difference-in-differences tables as in Table 7.1 in a transparent way that estimates a coherent causal effect (Cengiz et al., 2019). This type of aggregated DiD estimator (proposed by Callaway and Sant'Anna (2021) and Sun and Abraham (2021), among others, with some particulars and details varying across approaches) allows for heterogeneity across treatment time cohorts and is the current recommended approach for estimating policy impacts with DiD. This is in contrast to an older, regression-based approach that we recommend to avoid: fitting a model for each state and year with an indicator for whether the state had enacted RTC by that year and fixed effects for state and year.[11] This approach has been shown to estimate a non-interpretable weighted combination of individual two-by-two group comparisons, including some comparisons where both states are post-intervention (Goodman-Bacon, 2021). See Sun and Abraham (2021) and Callaway and Sant'Anna (2021) for further discussion and other forms of aggregation.

Figure 7.5b shows the overall "event study" plot: our aggregate estimate $\hat{\tau}_k^{\text{did}}$ along with 95% confidence intervals constructed via a bootstrap resampling procedure (discussed below). As in Section 7.3.2, we can use the estimates for years before the implementation ($k < 0$, left of the dashed lines) as a diagnostic check on the underlying parallel counterfactual trends assumption. Here the parallel trends assumption does not seem plausible. In the years leading up to enacting RTC, RTC states saw violent crime rates rising relative to their comparison states, either because violent crime was rising at a faster rate or because they were decreasing at a slower rate. Even though the aggregate DiD estimates point to a weakly significant increase in violent crime due to RTC, the event study plot in Figure 7.5b shows that this post-RTC difference is consistent with the difference in violent crime trends between RTC and comparison states during the pre-RTC period. Therefore, these estimates are not very credible because the key counterfactual parallel trends assumption is not plausible. To try to address this, we will use augmented SCM to estimate the cohort-level effects and then aggregate.

7.4.2.2 SCM with Staggered Adoption

As when aggregating from individual to time-cohort-level effect estimates with SCM estimators, there is a design choice on whether to find separate synthetic controls at the state or time-cohort level and then aggregate, or to find an aggregate synthetic control across states and time cohorts. Here there are methodological trade-offs due to the staggered adoption of RTC. Focusing on the average of the RTC states ensures that the violent crime rate for the average synthetic control will match that of the average RTC state; however, because states enacted RTC in different years, averaging in this way opens us to potential interpolation biases. If the underlying trends in the violent crime rate are changing substantially over time, averaging across time cohorts rather than *within* them can lead to bias. To circumvent this, Ben-Michael et al. (2022) propose a "partially pooled" approach that simultaneously finds a synthetic control that fits well for the average state and within individual states or time cohorts. In this chapter we will aggregate first to the time-cohort level, then partially pool across time cohorts.

Figure 7.6a shows the distribution of RMSPEs at the time-cohort level and for the average across time cohorts. For most time cohorts, the corresponding synthetic control fits reasonably well; the average time cohort has an RMSPE of 29.7 crimes per 100,000, and the four time cohorts (1987, 1991, 2001, 2014) with RMSPEs above 40 crimes per 100,000 each consist of a single state (FL, MT, MI, and IL, respectively). The fixed-effects augmented synthetic control for the average across time cohorts fits very well, with an RMSPE of only 4.7 crimes per 100,000. Figure 7.7a shows the "gap plot" for the average time cohort. As we expect from the RMSPE measure, the fit for the average across time cohorts is very good; on average, the synthetic controls closely match their respective time cohorts so that the average difference is near zero during the entire pre-RTC period.

7.4.2.3 Approaches to Inference

There are many possible approaches to inference with policy trial emulation. Our basic building block is to construct confidence intervals via re-sampling methods that re-sample entire states, such as the bootstrap, which are robust to correlations within a state over time. The results presented in this chapter

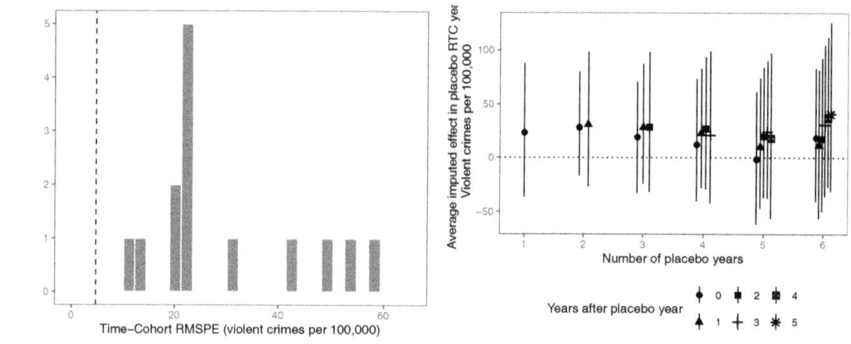

(a) (b)

FIGURE 7.6
(a) A Histogram of the Root Mean Square Prediction Error (RMSPE) for the Fixed-Effects Augmented Synthetic Control for Each Time Cohort. The dashed black line represents the RMSPE for the aggregated synthetic control. (b) In-Time Placebo Estimates and 95% Bootstrap Confidence Intervals of the Impact of RTC in a Held-Out Placebo RTC Adoption Year, setting the placebo adoption year to be one to seven years before RTC was enacted in each state.

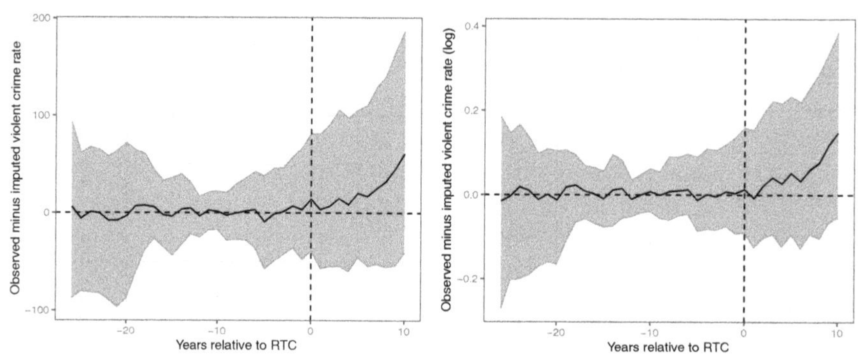

(a) (b)

FIGURE 7.7
An Aggregate "Gap Plot" Showing the Overall Fixed-Effects Augmented SCM Estimates of the Impact of RTC on the Violent Crime Rate in RTC States, along with 95% confidence intervals for (a) the violent crime rate per 100,000 and (b) the log violent crime rate per 100,000.

use the wild bootstrap procedure proposed by Otsu and Rai (2017) for matching estimators; see Ben-Michael et al. (2022) for discussion on implementing this approach for panel data methods and Callaway and Sant'Anna (2021) for related bootstrap-style approaches relying on efficient influence functions. Other approaches to inference are also available, including conformal inference approaches that provide exact, non-asymptotic confidence intervals (Chernozhukov et al., 2021; Cattaneo et al., 2021, 2025) (Figure 7.6).

When aggregating across multiple treatment time cohorts, an important consideration is that outcomes for the same state enter into the analysis in multiple places. Therefore, a key step to constructing confidence intervals is to track where and how outcomes for the same state enter into the estimate, and then to use an inference approach that relies on re-sampling states (or directly computing standard errors).

7.4.2.4 Placebo Checks and Impact Estimates

An important additional diagnostic is an out-of-sample check often called a placebo-in-time check, wherein we set the RTC year for each state to be earlier than it actually is. For example, if we set the RTC year to be two years earlier, this gives us two years of pre-RTC data for each RTC state that are not used to fit the synthetic controls but where the estimated "effect" of RTC should be zero. Figure 7.6b shows the average estimated effect in the placebo RTC year, as we vary the placebo year from one to seven years before its actual value for each state.[12] We find that while the placebo point estimates are not precisely zero, they average near 25 violent crimes per 100,000, and zero is well within the range of 95% confidence intervals, leading us to conclude that the synthetic control can reasonably impute the counterfactual violent crime rate.

These diagnostics and the event study plot in Figure 7.5b lead us to the conclusion that the fixed-effects augmented SCM estimates will be more credible than the DiD estimates. Figure 7.7a shows the estimates for the impact of RTC laws on the violent crime rate per 100,000 (as measured using UCR data). Overall, the point estimates are near zero for the first 5 years after enacting RTC, followed by an increase in the violent crime rate in the next 5 years. However, there is a substantial degree of uncertainty, and the confidence intervals contain zero in every time period.

7.4.2.5 Estimating Multiplicative Effects with the Log Violent Crime Rate

A benefit of the approach laid out in this chapter is that the key principles are agnostic to any transformations of the outcomes. So, for example, we can repeat the analysis above using the *log violent crime rate per 100,000* as our outcome of interest. Estimating impacts on the log scale allows us to measure multiplicative effects, which approximate the percent change in the violent crime rate relative to the no RTC baseline.[13] When evaluating impacts on transformed outcomes, one should conduct the same diagnostic and placebo checks as we have done above. In particular, for DiD the parallel trends assumption may seem plausible when evaluating the event study plot for some transformations of the outcomes, but not others. In the interest of space we do not report these diagnostics here – they are similar to the ones above. Figure 7.7b shows the estimated impact of RTC laws on the log scale. When looking at multiplicative, rather than additive effects, we estimate that there was an increase in the violent crime rate over time, with an average increase of roughly 15% ten years after enacting RTC, albeit with confidence intervals that include zero.

7.5 Additional Methodological Challenges and Future Directions

In this chapter we have considered the key components of operationalizing impact evaluations for gun policy research and estimating policy effects with modern statistical methods for causal inference. We have examined the key design decisions, assumptions underlying the analyses, and diagnostics for assessing these assumptions.

We have explicated these methods using a case study on evaluating the impact of right-to-carry laws. Overall, our difference-in-differences analysis showed evidence of pre-trends. The 10–20 years preceding the enactment of an RTC law saw the violent crime rate rising in RTC states relative to non-RTC states (and later adopters), casting doubt on the key identifying assumption of parallel counterfactual trends. Our fixed-effects augmented SCM analysis adjusted for these pre-trends, giving synthetic controls with good pre-treatment fit, and estimated null effects in the in-time placebo check. The estimated impacts of RTC on the violent crime rate are near zero, but noisy, though the estimated impact is larger 10 years after enacting RTC. When repeating the analysis to estimate multiplicative impacts by using the log violent crime rate as the outcome, the point estimates indicate that RTC led to an increase in violent crime over time, but there is insufficient evidence to reject a null hypothesis of no effect. The difference between the results with and without log-transforming the crime rate could be consistent with states with a lower baseline level of violent crime seeing a small absolute increase in violent crime that is still large relative to the baseline level. This result for the log violent crime rate mirrors the results in the original analysis in Donohue et al. (2019), which finds a 13–15% increase in violent crime. To contextualize these results, broadly researchers have found that RTC laws led to increases in some crime outcomes; for example, Doucette et al. (2023) found that shall-issue laws were associated with roughly a 10% increase in assaults with a gun. Other research has found that the adoption of these laws increased the rate of officer-involved shootings of civilians (Doucette et al., 2022) and has identified a link between shall-issue CCW law adoption and increases in firearm theft and decreases in homicide case closure rates (Donohue et al., 2023).

To conclude our discussion we will briefly touch on some of the additional methodological challenges that we have avoided in our development above because they constitute open problems in the field.

7.5.1 Estimating Impacts on Multiple Outcomes

In our analysis we have focused on measuring the impact of RTC laws on violent crime, but this is an aggregate of several distinct types of crime. We could instead have considered the impact on murder, rape, robbery, and aggravated assault separately, conducting a synthetic control analysis for each type of crime outcome. There are two related challenges with such an approach. First, finding synthetic controls separately for each crime outcome will lead to incompatible analyses. For instance, in Figure 7.4a we saw that the fixed-effects augmented synthetic control for West Virginia was composed of a combination of CO(50%), HI(35%), and NM (14%), but if we instead analyzed the murder rate, the augmented control would give a very different synthetic control, primarily weighting MA (36%), MO (36%), and HI(24%). If we were to analyze the four components of violent crime separately and then aggregate to estimate the impact on overall violent crime, we may get different estimates. In addition, if we are interested in subgroup effects (e.g., homicides stratified by the race of the victim), we would also need to check or ensure that the sum of the estimated impacts on the subgroups, weighted by the population, is equal to the estimated impact overall. Second, analyzing the outcomes separately precludes us from sharing information or structure across outcomes to get a better estimator. There has been some methodological work to understand how to perform combined analyses across multiple outcomes simultaneously (see, e.g., Sun et al., 2025), but optimally conducting such analyses remains an open question.

7.5.2 Measuring Outcomes at Different Temporal Resolutions

In our analyses we use violent crime measured annually; however, we could have used a finer temporal resolution, measuring crime rates at the quarterly or monthly level. Using quarterly or monthly crime rates would give us more pre-treatment time periods over which to fit our synthetic controls, which could lead to a better estimate. On the other hand, quarterly or monthly crime rates are noisier and can be subject to clear seasonal trends. Sun et al. (2024) show that the increase in noise often outweighs other benefits and a coarser aggregation can often be preferable. However, in some settings temporal aggregation can remove important signals in the data and lead to increased bias; the authors propose an approach that combines analyses at different temporal aggregations. Understanding and diagnosing when this harmful signal removal can occur is an open methodological question.

7.5.3 Accounting for the Structure of Count (and Often Rare) Outcomes

Most gun policy impact evaluations are focused on measuring the impact of the policy on counts, whether they be counts of crimes, deaths, injuries, etc. In contrast, the theoretical development for synthetic control and related methods has primarily been focused on continuous outcomes, so rates (e.g., the violent crime rate per 100,000) are often analyzed as continuous. For large jurisdictions this approximation is typically adequate, but it can fail when jurisdictions are small. This is particularly a problem for analyzing the impact of country or municipality-level policy changes on violent crime and especially homicides: many smaller cities and towns see just a handful of homicides per year, if any. Extending the approaches described in this chapter to properly account for such cases is another open direction for methodological work.

7.5.4 Accounting for Multiple Policies

As we discussed at the outset, there exist a wide variety of gun laws, each of which can have their own effect on gun violence. Moreover, different types of policies can interact with each other, and certain combinations of policies may have a larger effect on gun violence than either alone. Formalizing such interactions involves compound potential outcomes. For instance, to analyze the impact of RTC and stand-your-ground laws together, we would construct a two-level potential outcome $Y_{it}(s,g)$ representing the violent crime rate in state i in year t if the state enacted an RTC law in year s and a stand-your-ground law in year g; we could then consider counterfactual scenarios such as never enacting RTC but enacting a stand-your-ground law in a particular year. Such comparisons very quickly become unwieldy. However, without further (potentially quite strong) assumptions, it is not possible to avoid complex counterfactuals such as this. An important area of methodological research is to understand whether there are reasonable scenarios where the complexity is limited enough to estimate compound policy impacts, and if there are, whether there is sufficient data to represent the different variations in combinations of the policies, and then how to estimate those impacts.

7.5.5 Sensitivity Analysis and a Full Policy Evaluation Workflow

In this chapter we focused on designing and operationalizing a policy evaluation and estimating impacts after doing so. A final component to complete the policy evaluation "workflow" is to conduct a sensitivity analysis, inspecting how sensitive the results are to the underlying assumptions. Formally conducting such sensitivity analyses involves characterizing how assumptions can be violated and then mapping how estimates would change if the assumptions were to be violated in those ways. Approaches for doing so for the parallel trends assumption exist (see, e.g., Roth and Sant'Anna, 2023), but developing such procedures for estimation strategies, such as SCM, that do not rely on the parallel trends assumption is an open area of methodological exploration. Another form of sensitivity analysis is to consider sensitivity to data issues. For example, when using UCR data, there may be differential under-reporting of data across states and time that could bias effects estimated under either the DiD or SCM approach. Formalizing the potential for sensitivity to such bias is an open question. Finally, flexible and robust approaches to statistical inference remain an active area of research.

7.5.6 Qualitative and Mixed Methods Approaches

A final point relates to the role of qualitative and mixed methods research (e.g., McGinty et al., 2021). When conducting policy evaluations, there can be great value to using interviews and other qualitative or mixed methods approaches to gain information about what was happening "on the ground" in terms of the practical implementation of the policy and measurement and interpretation of the outcomes. This can help to identify other contemporaneous policies or unique events that would cast doubt on the counterfactual parallel trends assumption. It can also help to determine the true "start date" of a policy, especially if there is a gap between the legal enactment of a policy and its practical implementation. Such deep understanding can also help address data anomalies or strange patterns. We encourage quantitative researchers to engage with substantive experts to ensure the appropriate use of data and interpretation of study results, leading to more useful and accurate inferences about policy effects.

Notes

1. While most gun policies are at the state level, other policies or programs such as violence cessation programs or changes to law enforcement practices occur at lower jurisdictional levels, e.g., at the county or municipality level, in which case we would set the unit of analysis to be at this lower level. Furthermore, as we will see below, even when only dealing with states, we need not always set $N = 5$ and include all states (plus Washington DC) in our analysis.

2. In 2022, the US Supreme Court deemed the "proper cause" requirement unconstitutional in *New York State Rifle & Pistol Association v. Bruen*. Though the eight remaining states with may-issue laws have largely maintained the remainder of their policies, these laws are under increased scrutiny in the wake of this decision and may not be as restrictive as they were in the past.
3. A major exception is policy changes that are only possible due to other changes in law such as constitutional amendments or court decisions.
4. Inspecting the trend in the violent crime rate just prior to adopting the policy does not provide evidence for anticipation, as this assumption allows for correlation between the outcome and the adoption date (Athey and Imbens, 2022). For example, if we find that crime rates change just before adopting RTC, this could reflect an anticipatory effect of RTC, or it could reflect that states enact RTC in response to changing crime rates (which would be a violation of the counterfactual parallel trends assumption that we define below). Because it is difficult to disentangle the direction of causality, the assumption of no anticipation is difficult to test in general.
5. Note that unlike a typical statistical parameter, the effect for a single unit at a single event time τ_{ik} is a *random quantity* in this framework. Because of this, methods for statistical inference for the effect of a single unit often focus on constructing *prediction intervals* that include the inherent noise in the outcome. As we have defined it here, ATT_k is also a random quantity. Formally, some methods we describe below focus on estimating ATT_k while others target its expected value $\mathbb{E}[\text{ATT}_k]$. These procedures are typically analyzed under "large N" asymptotic settings where the number of units is growing, and so the difference $\text{ATT}_k - \mathbb{E}[\text{ATT}_k]$ is small. For the purposes of our discussion, we will not focus on this technical distinction.
6. There are many potential ways to define this assumption. To keep our development concise we focus on this particular definition. See Sun and Abraham (2021) and Callaway and Sant'Anna (2021) for further discussion.
7. So you should not do it.
8. Formally, SCM can estimate effects under a *latent factor* model that imposes a particular structure on the relationship between the violent crime rate across states at the same time, and over time within the same state. For more technical discussion, interested readers should consult Abadie et al. (2010), Arkhangelsky et al. (2021), and Ben-Michael et al. (2021, 2022a), among others.
9. Note that the event study plot in Figure 7.2b and the gap plot in Figure 7.3b are showing the same information, but for different estimators. In keeping with the nomenclature we will refer to these by separate names.
10. The DiD estimate in Equation (7.1) uses the last pre-RTC period as the comparison point. In contrast, Equation (7.5) uses all pre-RTC periods as comparison points and averages over these estimates.
11. That is, in R modeling notation, outcome ~ post_rtc + state + year. Note, though, that it is possible to reconstruct the DiD estimators we discuss via a regression with carefully constructed interaction terms (Wooldridge, 2021).
12. Beyond seven years there would be no "pre-treatment" information for the first treatment time cohort.

13. Formally, the difference in log potential outcomes is the log ratio of the potential outcomes: $\log Y_{iT_i+k}(T_i) - \log Y_{iT_i+k}(\infty) = \log \frac{Y_{iT_i+k}(T_i)}{Y_{iT_i+k}} \approx \frac{Y_{iT_i+k}(T_i) - Y_{iT_i+k}(\infty)}{Y_{iT_i+k}(\infty)}$ for percent changes between roughly -20% and 20%.

Bibliography

Abadie, A. (2021). Using Synthetic Controls: Feasibility, Data Requirements, and Methodological Aspects. *Journal of Economic Literature 59* (2), 391–425.

Abadie, A., A. Diamond, and J. Hainmueller (2010). Synthetic Control Methods for Comparative Case Studies: Estimating the Effect of California's Tobacco Control Program. *Journal of the American Statistical Association 105* (490), 493–505.

Abadie, A., A. Diamond, and J. Hainmueller (2015). Comparative Politics and the Synthetic Control Method. *American Journal of Political Science 59* (2), 495–510.

Abadie, A. and J. Gardeazabal (2003). The Economic Costs of Conflict: A Case Study of the Basque Country. *The American Economic Review 93* (1), 113–132.

Amjad, M., D. Shah, and D. Shen (2018). Robust Synthetic Control. *Journal of Machine Learning Research 19*, 1–51.

Angrist, J. D. and J.-S. Pischke (2008). *Mostly Harmless Econometrics: An Empiricist's Companion*. Princeton University Press.

Arkhangelsky, D., S. Athey, D. A. Hirshberg, G. W. Imbens, and S. Wager (2021). Synthetic Difference in Differences. *American Economic Review 111* (12), 4088–4118.

Asher, J. (2022, October). The FBI's Murder Rate Mystery. Vox. https://www.vox.com/policy-and-politics/2022/10/6/23389507/murder-crime-rate-fbi-report.

Athey, S., M. Bayati, N. Doudchenko, G. Imbens, and K. Khosravi (2021). Matrix Completion Methods for Causal Panel Data Models. *Journal of the American Statistical Association 116* (536), 1716–1730.

Athey, S. and G. W. Imbens (2022, January). Design-Based Analysis in Difference-In-Differences Settings with Staggered Adoption. *Journal of Econometrics 226* (1), 62–79.

Ault, A. (2021, August). Gun Violence Researchers Are Making Up for 20 Years of Lost Time. *JAMA 326* (8), 687.

Azad, H. A., M. C. Monuteaux, C. A. Rees, M. Siegel, R. Mannix, L. K. Lee, K. M. Sheehan, and E. W. Fleegler (2020, May). Child Access Prevention Firearm Laws and Firearm Fatalities Among Children Aged 0 to 14 Years, 1991–2016. *JAMA Pediatrics 174* (5), 463–469.

Bangalore, S. and F. H. Messerli (2013, October). Gun Ownership and Firearm-Related Deaths. *The American Journal of Medicine 126* (10), 873–876.

Banks, D., J. Hendrix, M. Hickman, and T. Kyckelhahn (2016). National Sources of Law Enforcement Employment Data. Technical Report Program Report NCJ 249681, U.S. Department of Justice, Office of Justice Programs.

Ben-Michael, E., A. Feller, and J. Rothstein (2021a). The Augmented Synthetic Control Method. *Journal of the American Statistical Association 116* (536), 1789–1803. arXiv: 1811.04170. Publisher: Taylor & Francis.

Ben-Michael, E., A. Feller, and E. A. Stuart (2021b). A Trial Emulation Approach for Policy Evaluations with Group-Level Longitudinal Data. *Epidemiology 32* (4), 533–540.

Ben-Michael, E., A. Feller, and J. Rothstein (2022). Synthetic Controls with Staggered Adoption. *Journal of the Royal Statistical Society. Series B: Statistical Methodology 84* (2), 351–381.

Callaway, B. and P. H. C. Sant'Anna (2021, December). Difference-in-Differences with Multiple Time Periods. *Journal of Econometrics 225* (2), 200–230.

Card, D. and A. B. Krueger (2000). Minimum Wages and Employment: A Case Study of the Fast-Food Industry in New Jersey and Pennsylvania: Reply. *The American Economic Review 90* (5), 1397–1420.

Castillo-Carniglia, A., R. M. C. Kagawa, M. Cerd´a, C. K. Crifasi, J. S. Vernick, D. W. Webster, and G. J. Wintemute (2019, February). California's Comprehensive Background Check and Misdemeanor Violence Prohibition Policies and Firearm Mortality. *Annals of Epidemiology 30*, 50–56.

Castillo-Carniglia, A., R. M. C. Kagawa, D. W. Webster, J. S. Vernick, M. Cerd´a, and G. J. Wintemute (2018, December). Comprehensive Background Check Policy and Firearm Background Checks in Three US States. *Injury Prevention 24* (6), 431–436.

Cattaneo, M. D., Y. Feng, F. Palomba, and R. Titiunik (2025). Uncertainty Quantification in Synthetic Controls with Staggered Treatment Adoption. *Review of Economics and Statistics 107* (1), 1–46.

Cattaneo, M. D., Y. Feng, and R. Titiunik (2021). Prediction Intervals for Synthetic Control Methods. *Journal of the American Statistical Association 116* (536), 1865–1880.

Cengiz, D., A. Dube, A. Lindner, and B. Zipperer (2019). The Effect of Minimum Wages on Low-Wage Jobs. *The Quarterly Journal of Economics*, 1405–1454.

Chernozhukov, V., K. Wu¨thrich, and Y. Zhu (2021). An Exact and Robust Conformal Inference Method for Counterfactual and Synthetic Controls. *Journal of the American Statistical Association 116* (536), 1849–1864.

Crifasi, C. K., M. Merrill-Francis, A. McCourt, J. S. Vernick, G. J. Wintemute, and D. W. Webster (2018, June). Association between Firearm Laws and Homicide in Urban Counties. *Journal of Urban Health 95* (3), 383–390.

Crifasi, C. K., J. S. Meyers, J. S. Vernick, and D. W. Webster (2015, October). Effects of Changes in Permit-to-Purchase Handgun Laws in Connecticut and Missouri on Suicide Rates. *Preventive Medicine 79*, 43–49.

Danaei, G., L. A. García Rodríguez, O. F. Cantero, R. W. Logan, and M. A. Hern´an (2018). Electronic Medical Records Can Be Used to Emulate Target Trials of Sustained Treatment Strategies. *Journal of Clinical Epidemiology 96*, 12–22.

Davis, A., R. Kim, and C. Crifasi (2023). U.S. Gun Violence in 2021: An Accounting of a Public Health Crisis. Technical report, Johns Hopkins Center for Gun Violence Solutions. The Johns Hopkins Bloomberg School of Public Health.

Degli Esposti, M., D. J. Wiebe, A. Gasparrini, and D. K. Humphreys (2022, February). Analysis of "Stand Your Ground" Self-Defense Laws and Statewide Rates of Homicides and Firearm Homicides. *JAMA Network Open 5* (2), e220077.

Dickerman, B. A., X. García-Albéniz, R. W. Logan, S. Denaxas, and M. A. Hern´an (2019). Avoidable Flaws in Observational Analyses: An Application to Statins and Cancer. *Nature Medicine 25* (10), 1601–1606.

Donohue, J. J., A. Aneja, and K. D. Weber (2019). Right-to-carry Laws and Violent Crime: A Comprehensive Assessment Using Panel Data and a State-Level Synthetic Control Analysis. *Journal of Empirical Legal Studies 16* (2), 198–247.

Donohue, J. J., S. V. Cai, M. V. Bondy, and P. J. Cook (2022, June). Why Does Right-to-Carry Cause Violent Crime to Increase? National Bureau of Economic Research, Working Paper No. 30190. https://www.nber.org/papers/w30190.

Doucette, M. L., C. K. Crifasi, and S. Frattaroli (2019, December). Right-to-Carry Laws and Firearm Workplace Homicides: A Longitudinal Analysis (1992–2017). *American Journal of Public Health 109* (12), 1747–1753.

Doucette, M. L., C. K. Crifasi, A. D. McCourt, J. A. Ward, R. L. Fix, and D. W. Webster (2023a). Deregulation of Public Civilian Gun Carrying and Violent Crimes: A Longitudinal Analysis 1981–2019. *Criminology & Public Policy n/a* 23 (4), 833–861.

Doucette, M. L., C. Green, J. Necci Dineen, D. Shapiro, and K. M. Raissian (2021, October). Impact of ShotSpotter Technology on Firearm Homicides and Arrests Among Large Metropolitan Counties: A Longitudinal Analysis, 1999–2016. *Journal of Urban Health 98* (5), 609–621.

Doucette, M. L., A. D. McCourt, C. K. Crifasi, and D. W. Webster (2023b). Impact of Changes to Concealed-Carry Weapons Laws on Fatal and Nonfatal Violent Crime, 1980–2019. *American Journal of Epidemiology 192* (3), 342–355.

Doucette, M. L., J. A. Ward, A. D. McCourt, D. Webster, and C. K. Crifasi (2022, June). Officer-Involved Shootings and Concealed Carry Weapons Permitting Laws: Analysis of Gun Violence Archive Data, 2014–2020. *Journal of Urban Health 99* (3), 373–384.

Doudchenko, N. and G. W. Imbens (2016, October). Balancing, Regression, Difference-In-Differences and Synthetic Control Methods: A Synthesis. National Bureau of Economic Research, Working Paper No. 22791. https://www.nber.org/papers/w22791.

Ferman, B. and C. Pinto (2021). Synthetic Controls with Imperfect Pre-treatment Fit. *Quantitative Economics 12(4), 1197–1221.*

Goodman-Bacon, A. (2021, December). Difference-in-differences with Variation in Treatment Timing. *Journal of Econometrics 225* (2), 254–277.

Hartman, E. and F. D. Hidalgo (2018). An Equivalence Approach to Balance and Placebo Tests. *American Journal of Political Science 62* (4), 1000–1013.

Hasegawa, R. B., S. K. Deshpande, D. S. Small, and P. R. Rosenbaum (2020, August). Causal Inference with Two Versions of Treatment. *Journal of Educational and Behavioral Statistics 45* (4), 426–445.

Hasegawa, R. B., D. W. Webster, and D. S. Small (2019, May). Evaluating Missouri's Handgun Purchaser Law: A Bracketing Method for Addressing Concerns About History Interacting with Group. *Epidemiology 30* (3), 371.

Hepburn, L., D. Azrael, M. Miller, and D. Hemenway (2006, August). The Effect of Child Access Prevention Laws on Unintentional Child Firearm Fatalities, 1979–2000. *Journal of Trauma and Acute Care Surgery 61* (2), 423.

Kagawa, R., A. Charbonneau, C. McCort, A. McCourt, J. Vernick, D. Webster, and G. Wintemute (2023, April). Effects of Comprehensive Background-Check Policies on Firearm Fatalities in 4 States. *American Journal of Epidemiology 192* (4), 539–548.

Kagawa, R. M. C., A. Castillo-Carniglia, J. S. Vernick, D. Webster, C. Crifasi, K. E. Rudolph, M. Cerd´a, A. Shev, and G. J. Wintemute (2018, July). Repeal of Comprehensive Background Check Policies and Firearm Homicide and Suicide. *Epidemiology 29* (4), 494.

Knopov, A., M. Siegel, Z. Xuan, E. F. Rothman, S. W. Cronin, and D. Hemenway (2019, October). The Impact of State Firearm Laws on Homicide Rates among Black and White Populations in the United States, 1991–2016. *Health & Social Work 44* (4), 232–240.

Lantz, B. (2022, June). Why More Agencies and Researchers Should Embrace the Upcoming NIBRS Transition: Contributions and Promise of the NIBRS Data. *American Journal of Criminal Justice 47* (3), 462–484.

Li, W. (2022, June). What Can FBI Data Say About Crime in 2021? It's Too Unreliable to Tell. *The Marshall Project*.

Li, W., A. R. Calder´on, D. Eads, and A. Flagg (2023, August). See If Police in Your State Reported Crime Data to the FBI. Section: Analysis. *The Marshall Project*.

Lott, J. R. and D. B. Mustard (1997, January). Crime, Deterrence, and Right-to-Carry Concealed Handguns. *The Journal of Legal Studies 26* (1), 1–68.

Maltz, M. D. (2006). Analysis of Missingness in UCR Crime Data. Technical report, National Institute of Justice.

McCourt, A. D., C. K. Crifasi, E. A. Stuart, J. S. Vernick, R. M. Kagawa, G. J. Wintemute, and D. W. Webster (2020, October). Purchaser Licensing, Point-of-Sale Background Check Laws, and Firearm Homicide and Suicide in 4 US States, 1985–2017. *American Journal of Public Health 110* (10), 1546–1552.

McElroy, M. B. and W. Wang (2017, June). Seemingly Inextricable Dynamic Differences: The Case of Concealed Gun Permit, *Violent Crime and State Panel Data*. https://papers.ssrn.com/sol3/papers.cfm?abstract_id=2992058

McGinty, E. E., K. N. Tormohlen, C. L. Barry, M. C. Bicket, L. Rutkow, and E. A. Stuart (2021, January). Protocol: Mixed-Methods Study of How Implementation of US State Medical Cannabis Laws Affects Treatment of Chronic Non-cancer Pain and Adverse Opioid Outcomes. *Implementation Science: IS 16* (1), 2.

Neyman, J. (1990 [1923]). On the Application of Probability Theory to Agricultural Experiments. Essay on Principles. Section 9. *Statistical Science 5* (4), 465–472.

Otsu, T. and Y. Rai (2017). Bootstrap Inference of Matching Estimators for Average Treatment Effects. *Journal of the American Statistical Association 112*, 1720–1732.

Parker, S. T. (2022, May). Measuring Gun Violence in Police Data Sources: Transitioning to NIBRS. *Injury Epidemiology 9* (1), 15.

Roth, J. and P. H. C. Sant'Anna (2023). When Is Parallel Trends Sensitive to Functional Form? *Econometrica 91* (2), 737–747.

Roth, J., P. H. C. Sant'Anna, A. Bilinski, and J. Poe (2023, August). What's Trending in Difference-in-differences? A Synthesis of the Recent Econometrics Literature. *Journal of Econometrics 235* (2), 2218–2244.

Rowhani-Rahbar, A., M. A. Bellenger, and F. P. Rivara (2019, November). Firearm Violence Research: Improving Availability, Accessibility, and Content of Firearm-Related Data Systems. *JAMA 322* (19), 1857–1858.

Rubin, D. B. (1974). Estimating Causal Effects of Treatments in Randomized and Nonrandomized Studies. *Journal of Educational Psychology 66* (5), 688.

Rubin, D. B. (1980). Randomization Analysis of Experimental Data: The Fisher Randomization Test Comment. *Journal of the American Statistical Association 75* (371), 591–593.

Schell, T. L., M. Cefalu, B. A. Griffin, R. Smart, and A. R. Morral (2020, June). Changes in Firearm Mortality Following the Implementation of State Laws Regulating Firearm Access and Use. *Proceedings of the National Academy of Sciences 117* (26), 14906–14910.

Siegel, M., M. Pahn, Z. Xuan, E. Fleegler, and D. Hemenway (2019, October). The Impact of State Firearm Laws on Homicide and Suicide Deaths in the USA, 1991–2016: A Panel Study. *Journal of General Internal Medicine 34* (10), 2021–2028.

Siegel, M., M. Pahn, Z. Xuan, C. S. Ross, S. Galea, B. Kalesan, E. Fleegler, and K. A. Goss (2017, July). Firearm-Related Laws in All 50 US States, 1991–2016. *American Journal of Public Health 107* (7), 1122–1129.

Smart, R., A. R. Morral, R. Ramchand, A. Charbonneau, J. Williams, S. Smucker, S. Cherney, and L. Xenakis (2023, January). The Science of Gun Policy: A Critical Synthesis of Research Evidence on the Effects of Gun Policies in the United States, Third Edition. Technical report, RAND Corporation.

Stark, D. E. and N. H. Shah (2017, January). Funding and Publication of Research on Gun Violence and Other Leading Causes of Death. *JAMA 317* (1), 84–85.

Sun, L. and S. Abraham (2021, December). Estimating Dynamic Treatment Effects in Event Studies with Heterogeneous Treatment Effects. *Journal of Econometrics 225* (2), 175–199.

Sun, L., E. Ben-Michael, and A. Feller (2024). Temporal Aggregation for the Synthetic Control Method. *AEA Papers and Proceedings 114*, 614–617.

Sun, L., E. Ben-Michael, and A. Feller (2025). Using Multiple Outcomes to Improve the Synthetic Control Method. *Review of Economics and Statistics*. https://doi.org/10.1162/rest_a_01592.

Van Der Wal, W. M. (2022, December). Marginal Structural Models to Estimate Causal Effects of Right-to-Carry Laws on Crime. *Statistics and Public Policy 9* (1), 163–174.

Vernick, J. S., T. Alcorn, and J. Horwitz (2017, April). Background Checks for all Gun Buyers and Gun Violence Restraining Orders: State Efforts to Keep Guns from High-Risk Persons. *Journal of Law, Medicine & Ethics 45* (S1), 98–102.

Webster, D., C. K. Crifasi, and J. S. Vernick (2014, April). Effects of the Repeal of Missouri's Handgun Purchaser Licensing Law on Homicides. *Journal of Urban Health 91* (2), 293–302.

Webster, D. W., A. D. McCourt, C. K. Crifasi, M. D. Booty, and E. A. Stuart (2020). Evidence Concerning the Regulation of Firearms Design, Sale, and Carrying on Fatal Mass Shootings in the United States. *Criminology & Public Policy 19* (1), 171–212.

Webster, D. W., J. S. Vernick, J. Ludwig, and K. J. Lester (1997, June). Flawed Gun Policy Research Could Endanger Public Safety. *American Journal of Public Health 87* (6), 918–921.

Webster, D. W., J. S. Vernick, A. M. Zeoli, and J. A. Manganello (2004, August). Association between Youth-Focused Firearm Laws and Youth Suicides. *JAMA 292* (5), 594–601.

Wooldridge, J. M. (2021). Two-Way Fixed Effects, the Two-Way Mundlak Regression, and Difference-in-Differences Estimators. Available at SSRN: https://ssrn.com/abstract=3906345 or http://dx.doi.org/10.2139/ssrn.3906345.

Xu, Y. (2017). Generalized Synthetic Control Method: Causal Inference with Interactive Fixed Effects Models. *Political Analysis 25* (1), 57–76.

Yakubovich, A. R., M. D. Esposti, B. C. L. Lange, G. J. Melendez-Torres, A. Parmar, D. J. Wiebe, and D. K. Humphreys (2021, April). Effects of Laws Expanding Civilian Rights to Use Deadly Force in Self-Defense on Violence and Crime: A Systematic Review. *American Journal of Public Health 111* (4), e1–e14.

8

Scaling Hawkes Processes

Seyoon Ko, Jasen Zhang, and Andrew J. Holbrook

8.1 Introduction

Hawkes processes (HP) model event contagion. As such, this large class of models finds use in the analysis of earthquakes (Hawkes, 1973; Ogata, 1988; Zhuang et al., 2004), viral memes (Yang and Zha, 2013; Mei and Eisner, 2017), biological neural activity (Linderman and Adams, 2014; Truccolo, 2016; Linderman et al., 2017), viral epidemics (Kim, 2011; Meyer et al., 2014; Choi et al., 2015; Rizoiu et al., 2018; Kelly et al., 2019; Holbrook et al., 2022a, b), and financial transactions (Embrechts et al., 2011; Chavez-Demoulin and McGill, 2012; Hardiman et al., 2013; Hawkes, 2018). In all these applications, the observation of an event (whatever it may be) at a specific time increases the probability of observing another event soon after.

In addition to dates or timestamps, individual data may contain spatial coordinates, categorical assignments, or some other information about the events they describe. Researchers have developed different kinds of HP to account for different kinds of data. Applying an HP often involves fitting the model to data observed. The exact method one should use to fit a specific HP model depends on the kind of model, the size of the data, the purpose of the analysis, the computational resources at hand, and the amount of time one would require to implement the method. A streamlined introduction, Section 8.2 provides examples involving five methods for fitting HP of various forms, paying attention to each method's time complexity as a function of data count and relevant dimensions. Such computational considerations are paramount in the age of big data, especially since many different approaches to fitting various HP to N observed events typically require a quadratic $O(N^2)$ number of floating point operations. Such computational burdens prevail regardless of data application, but scientists working in

DOI: 10.1201/9781003430773-9

fields featuring massive quantities of events must implement their HP models with scalability in mind.

As a primary example, HPs have emerged as statistical tools enabling the use of data to mathematically characterize the contagiousness of violence, a premise advanced within the criminology literature (Loftin, 1986; Fagan et al., 2007; Taylor et al., 2013). Mohler (2014) develops a novel spatiotemporal HP (StHP) model for homicides conditioned on other violent crimes. Mohler (2014) analyzes subsets of 80,000 events in Chicago from 2007 to 2012 using an EM algorithm (cf., Section 8.2, Example 3), requiring roughly 200 iterations for an analysis involving 3-year subsets of the data. For each subset, homicides are on the order of thousands while total violent crimes are in the tens-of-thousands; across all EM iterations, the total number of floating point operations is $O(10^9)$, reportedly requiring one hour of time on a personal laptop. Linderman and Adams (2014) use the multivariate HP (MHP) framework described in Section 8.2, Example 4, to analyze over 1,600 gang-related homicides across 77 Chicago neighborhoods from 1980 to 1995. The authors' careful GPU implementation achieves 5–50 Gibbs updates per second across all examples considered, and it is not clear exactly where their Chicago gang-violence analysis lands on this interval. The authors do not seem to report the number of Gibbs updates required, but it is reasonable to assume that they use orders-of-magnitude more than the number of EM updates required to fit the albeit smaller model of Mohler (2014).

Loeffler and Flaxman (2018) use an StHP to analyze over 9,000 gunshots recorded by an acoustic gunshot locator system (AGLS) in Washington DC from 2010 to 2012. After specifying prior distributions on StHP parameters, Loeffler and Flaxman (2018) perform Bayesian inference using Hamiltonian Monte Carlo with the help of the Stan probabilistic programming language. In total, the authors simulate four MCMC chains for 200 iterations each within an unspecified amount of time. For convergence diagnostics, they do not report effective sample sizes (ESS) but do report a Gelman-Rubin $\hat{R}$ statistic close to 1. Again working within the Bayesian framework, Holbrook et al. (2021b) scale a similar StHP model to over 85,000 gunshots recorded by the same AGLS in Washington DC from 2006 to 2019 with the help of hand-coded GPU and multi-core CPU routines for the StHP likelihood. The need for such parallelization arises from the fact that a 10-fold increase in events results in a 100-fold increase in likelihood calculation floating point operations. Holbrook et al. (2021a) implement their fast likelihood calculations within an adaptive Metropolis-Hastings (MH) algorithm (Metropolis et al., 1953; Hastings, 1970), generating four Markov chain Monte Carlo (MCMC) chains of length 10,000, each requiring 10 hours and obtaining a minimum ESS of 1,700. Using the same fast likelihood calculations within a similar adaptive MH scheme, Holbrook et al. (2022a) extend this model to include event-specific latent locations with prior probabilities reflecting the spatial precision of location measurements. Although such a data augmentation strategy improves model fit in simulations, the resulting $O(N)$ increase in model parameters restricts applicability to events numbered in the low thousands.

Recently, Ko et al. (2024) introduce a multi-GPU, high-performance computing strategy for fitting Bayesian StHP models to big spatiotemporal data using the adaptive MH algorithms at the core of Holbrook et al. (2021, 2022a). Ko et al. (2024) apply their computational inference framework to the analysis of roughly one million COVID-19 cases in the US; in Section 8.3, we show that this framework readily translates to gun violence data. This work is mainly computational in nature, and we do not discuss other contemporaneous contributions in this area that improve the interpretability of HP analyses, expand the expressivity of HP models, or enable more effective forecasting from HP models once fit (Sha et al., 2020; Park et al., 2021; Brantingham et al., 2021; Zhu and Xie, 2022; Dong et al., 2023). We do, however, note that the general fine-grained parallelization strategies for HP likelihood calculations are largely applicable to many of these works and hope that the computational contributions we consider serve as a road map for scaling ever more interpretable, flexible, and complex HP models to big data. Before entering in on the analysis of big American gun violence data, we review several types of HP and the computational complexity of strategies for fitting them.

8.2 Strategies for Fitting Hawkes Processes

An HP is a stochastic point process model that exhibits self-exciting behavior (Hawkes, 1971a, b, 1972). In its basic form, an HP characterizes collections of random events in time $\{t_1, t_2, \ldots\} \in [0, \infty)$ using a conditional intensity function

$$\lambda^*(t) = \lambda(t \mid t_n < t) = \lambda_0 + \sum_{t_n < t} \xi(t - t_n) \tag{8.1}$$

that describes the instantaneous rate of event arrival at time t given preceding event times. Here, $\lambda_0 > 0$ is a fixed background rate and $\xi(\cdot) > 0$ is the excitation function or triggering kernel, the positivity of which increases $\lambda^*(t)$ immediately after each event t_n. This increase in the arrival rate may correspond to the occurrence of more events, each of which provides its own contribution to $\lambda^*(t)$. Hawkes and Oakes (1974) relate the conditional intensity-based representation (8.1) to an immigrant-offspring process representation that starts with a generation of immigrant events $\mathcal{G}_0 = \{t_{0,1}, t_{0,2}, \ldots\}$ following a homogeneous Poisson process with rate λ_0. Each background event $t_{0,n}$ gives birth to $O_{1,n} \sim \text{Poisson}(\beta)$ offspring events, where β is the branching number

$$\beta = \int_{t_{0,n}}^{\infty} \xi(t - t_{0,n})dt = \int_{0}^{\infty} \xi(t)dt. \tag{8.2}$$

If $O_{1,n} > 0$, the $O_{1,n}$ offspring occur at times $\mathcal{G}_{1,n} = \{t_{1,n,1}, \ldots t_{1,n,O_{1,n}}\} \overset{iid}{\sim} \xi(t - t_{0,n}) / \beta$. The second generation is the union $\mathcal{G}_1 = \cup_{n \in \mathcal{G}_0} \mathcal{G}_{1,n} = \{t_{1,1} \ldots t_{1,O_1}\}$, where $O_1 = \sum_n O_{1,n}$, and it is convenient to drop the index relating a child to its parent. The process continues in perpetuity, with each generation $\mathcal{G}_{g+1}$ consisting of offspring from members of the previous generation $\mathcal{G}_g$ and the entire HP sample path being the union $\cup_g \mathcal{G}_g$.

Myriad extensions to the basic HP exist and take forms adapted to the modeling of data arising from various phenomena. Both the conditional intensity and the immigrant-offspring representation continue to apply for many of these modeling extensions. As is common in many areas of statistics, one may fit an HP model to observed data using the method of moments or likelihood-informed methods such as maximum likelihood, EM, or Bayesian inference. Depending on model specifics and the technique one uses to fit one's model, it may be more convenient (computationally or otherwise) to use one of the HP representations or the other. This section is not a comprehensive review of the HP zoo and its applications. The reader may consult Laub et al. (2015, 2021) for more comprehensive reviews, Hawkes (2018) for a review of the HP in finance, and Reinhart (2018) for a review of the spatiotemporal HP. Here, we quickly touch on a small number of HP models and briefly discuss a few methods for fitting these models, paying attention to methods' relative benefits and the quadratic $O(N^2)$ time-complexity required to fit most interesting HP models to N observations, regardless of representation used.

Example 1 (Method of Moments). *Da Fonseca and Zaatour (2014) consider a basic exponential HP with rate function*

$$\lambda^*(t) = \lambda + (\lambda_0 - \lambda)e^{-\omega t} + \xi_0 \sum_{t_n < t} e^{-\omega(t - t_n)} \tag{8.3}$$

and derive the asymptotic moments for the number of jumps over intervals of length τ. *Letting* N_t *be the number of events on the interval* $[0, t]$, *these moments include*

$$\mu_{\tau,1} = \lim_{t \to \infty} \mathbb{E}(N_{t+\tau} - N_t | \lambda_0) = \frac{\lambda\omega}{\omega - \xi_0}\tau \tag{8.4}$$

$$\mu_{\tau,2} = \lim_{t\to\infty}\mathbb{E}\left(\left(N_{t+\tau}-N_t\right)^2|\lambda_0\right)-\mu_{\tau,1}^2 = \frac{\lambda\omega}{\omega-\xi_0}\left(\frac{\tau\omega^2}{\left(\omega-\xi_0\right)^2}+\left(1-\frac{\omega^2}{\left(\omega-\xi_0\right)^2}\right)\frac{1-e^{(\xi_0-\omega)\tau}}{\omega-\xi_0}\right)$$

as well as the autocovariance between counts of intervals separated by arbitrary lags. One may obtain corresponding empirical moments $\hat{\mu}_{\tau,1}$ *and* $\hat{\mu}_{\tau,2}$ *by dividing the data observation interval* $[t_0,\ T]$ *into B bins of length* τ *and summing over bin-wise event counts* C_b*:*

$$\hat{\mu}_{\tau,1} = \frac{1}{B}\sum_{b=1}^{B} C_b,\ \hat{\mu}_{\tau,2} = \frac{1}{B}\sum_{b=1}^{B} C_b^2 - \hat{\mu}_{\tau,1}^2. \tag{8.5}$$

Here, one may specify t_0 *to discard a certain number of early events as burn-in while ensuring that* τ *divides the observation interval. Da Fonseca and Zaatour (2014) use the generalized method of moments (Hall, 2005) to perform moment matching by defining* $\theta = (\lambda, \xi_0, \omega)$*, letting* $\hat{\mu}$ *and* $\mu(\cdot)$ *be vectors of empirical and theoretical moments, respectively, and using Levenberg-Marquardt (Levenberg, 1944; Marquardt, 1963) to find*

$$\hat{\theta} = \underset{\theta}{argmin}\left(\hat{\mu}_\tau - \mu_\tau(\theta)\right)^T\left(\hat{\mu}_\tau - \mu_\tau(\theta)\right).$$

In fact, any nonlinear optimization routine will do, and the generalized method of moments provides consistent estimators that satisfy asymptotic normality. In terms of empirical estimator accuracy and variability, Da Fonseca and Zaatour (2014) show that their method can perform better or worse than maximum likelihood (discussed below) depending on the moments used. They do not communicate speedups over the latter but note that computing for their method is "negligible" and that even the fast maximum likelihood method of Ozaki (1979) "takes a few minutes."

A drawback for the method of moments is that one must know theoretical moments in closed form. While this may be possible for simple models such as the exponential HP, the method of moments does not provide a recipe for scalably robust data science involving repeated application of multiple competing models within a single scientific workflow. What it gains in its comparably fast $O(N)$ time-complexity, it loses in person-hours since each new model requires erudite derivations, examples of which are provided in the impressive work of Daw and Pender (2023). On the other hand, maximum likelihood provides a more widely applicable method for fitting HP but generally requires $O(N^2)$ floating-point operations for each gradient evaluation, with Ozaki (1979) providing one prominent exception.

Example 2 (Maximum Likelihood). *For a point process with conditional intensity* $\lambda^*(t)$ *parameterized by* θ*, the likelihood of observing events* $\{t_1,\ldots,t_N\}$ *on an observation interval* $[0,T]$ *is (Rubin, 1972; Daley and Vere-Jones, 2003)*

$$L\left(\{t_n\}_{n=1}^{N} \mid \boldsymbol{\theta}, T\right) = e^{-\int_0^T \lambda^*(t)dt} \prod_{n=1}^{N} \lambda^*(t_n) = e^{-\Lambda(T)} \prod_{n=1}^{N} \lambda^*(t_n), \tag{8.6}$$

and the log-likelihood for the HP with conditional intensity (8.1) *is*

$$\ell\left(\{t_n\}_{n=1}^{N} \mid \boldsymbol{\theta}, T\right) = -\Lambda(T) + \sum_{n=1}^{N} \log\left(\lambda_0 + \xi_0 \sum_{t_{n'} < t_n} \xi(t_n - t_{n'})\right). \tag{8.7}$$

For the simplified version of the exponential HP conditional intensity (8.3)

$$\lambda^*(t) = \lambda_0 + \xi_0 \sum_{t_n < t} e^{-\omega(t - t_n)}, \tag{8.8}$$

the log-likelihood is

$$\ell\left(\{t_n\}_{n=1}^{N} \mid \boldsymbol{\theta}, t_N\right) = -\lambda_0 t_N + \sum_{n=1}^{N} \left(\log\left(\lambda_0 + \xi_0 \sum_{n'=1}^{n-1} e^{-\omega(t_n - t_{n'})}\right) + \frac{\xi_0}{\omega}\left(e^{-\omega(t_N - t_n)} - 1\right)\right), \tag{8.9}$$

where we follow common practice and take t_N as endpoint of the observation interval. While (8.9) *and its double summation ostensibly suffer $O(N^2)$ time-complexity, a recursion reveals linear complexity in N (Ozaki, 1979; Laub et al., 2015). For $n = 2, \dots, N$, let $\psi(n) = \sum_{n'=1}^{n-1} e^{-\omega(t_n - t_{n'})}$. Then*

$$\begin{aligned}\psi(n) &= e^{-\omega t_n + \omega t_{n-1}} \sum_{n'=1}^{n-1} e^{-\omega t_{n-1} + \omega t_{n'}} = e^{-\omega(t_n - t_{n-1})}\left(1 + \sum_{n'=1}^{n-2} e^{-\omega t_{n-1} - \omega t_{n'}}\right) \\ &= e^{-\omega(t_n - t_{n-1})}\left(1 + \psi(n-1)\right).\end{aligned}$$

Letting $\psi(1) = 0$, the log-likelihood is

$$\ell\left(\{t_n\}_{n=1}^{N} \mid \boldsymbol{\theta}, t_N\right) = -\lambda_0 t_N + \sum_{n=1}^{N} \left(\log\left(\lambda_0 + \xi_0 \psi(n)\right) + \frac{\xi_0}{\omega}\left(e^{-\omega(t_N - t_n)} - 1\right)\right).$$

Through similar recursions, the log-likelihood gradient and Hessian yield $O(N)$-time forms that one may implement within one's favorite nonlinear optimization algorithm.

Likelihood-based methods flexibly adapt to HP model extensions but pay the price in their $O(N^2)$ time-complexity. Log-likelihood formula (8.7) continues to apply if one specifies the power-law kernel $\xi(t) = \xi_0 \omega / (1 + \omega t)^{p+1}$, but no known recursion facilitates similar $O(N)$ computations. Beyond simple changes to $\xi(\cdot)$, likelihoods are available for more complex processes. A basic spatiotemporal HP (StHP) models the usual temporal event data t_n

along with corresponding locations $\mathbf{x}_n$ within some spatial domain $\mathcal{X}$ using the conditional intensity

$$\lambda^*(t,x)=\lambda_0(\mathbf{x})+\sum_{t_n<t}\xi(t-t_n,\mathbf{x}-\mathbf{x}_n) \tag{8.10}$$

for $\lambda_0(\cdot),\xi(\cdot,\cdot)\geq 0$, and the likelihood for a general StHP takes the recognizable form (Daley and Vere-Jones, 2003)

$$L\left(\{(t_n,\mathbf{x}_n)\}_{n=1}^{N}\mid\boldsymbol{\theta},T\right)=e^{-\int_{\mathcal{X}}\int_0^T\lambda^*(t,\mathbf{x})dtd\mathbf{x}}\prod_{n=1}^{N}\lambda^*(t_n,\mathbf{x}_n)=e^{-\Lambda(T,\mathcal{X})}\prod_{n=1}^{N}\lambda^*(t_n,\mathbf{x}_n). \tag{8.11}$$

Unfortunately, Veen and Schoenberg (2008) show that the log-likelihood can be relatively flat around its maximum for some important StHP and that this leads to convergence issues for popular numerical optimization algorithms such as Newton-Raphson.

Example 3 (Expectation Maximization). *Relying on the immigrant-offspring representation of the HP, Veen and Schoenberg (2008) propose augmenting each data point $(t_n,\mathbf{x}_n)$ for $n\in\{1,\dots,N\}$ with an unobserved latent variable $z_n\in\{0,1,\dots,n-1\}$ that indicates the parent of event n ($z_n=0$ for n an immigrant event) and using the EM algorithm (Dempster et al., 1977) to maximize the likelihood*

$$L\left(\{(t_n,\mathbf{x}_n)\}_{n=1}^{N}\mid\boldsymbol{\theta},T\right)=\sum_{z_1,\dots,z_N}L\left(\{(t_n,\mathbf{x}_n,z_n)\}_{n=1}^{N}\mid\boldsymbol{\theta},T\right) \tag{8.12}$$

obtained by marginalizing the complete-data likelihood

$$L\left(\{(t_n,\mathbf{x}_n.z_n)\}_{n=1}^{N}\mid\boldsymbol{\theta},T\right)\propto\prod_{n=1}^{N}\lambda_0(\mathbf{x}_n)^{\mathcal{I}_{[z_n=0]}}\prod_{n'=1}^{n-1}\xi(t_n-t_{n'},\mathbf{x}_n-\mathbf{x}_{n'})^{\mathcal{I}_{[z_n=n']}},$$

where $\mathcal{I}_{[\cdot]}$ is the indicator function. The EM algorithm of Veen and Schoenberg (2008) begins with some parameter value $\theta^{(0)}$ and alternates between expectation and maximization steps, performance of which together contribute to a single iteration. At iteration s, the E-step uses parameter $\theta^{(s-1)}$ to obtain

$$\Pr(z_n=n')=\mathbb{E}_{z_n}\left(\mathcal{I}_{[z_n=n']}\right)=\begin{cases}\dfrac{\xi(t_n-t_{n'},\mathbf{x}_n-\mathbf{x}_{n'})}{\lambda^*(t_n,\mathbf{x}_n)} & n'\in\{1,\dots,n-1\}\\[2ex] \dfrac{\lambda_0(\mathbf{x}_n)}{\lambda^*(t_n,\mathbf{x}_n)} & n'=0\end{cases} \tag{8.13}$$

for $n\in\{1,\dots,N\}$. The M-step then maximizes the expected complete-data log-likelihood

$$\mathbb{E}_{z_1,\ldots,z_N}\ell\left(\left\{(t_n,\mathbf{x}_n)\right\}_{n=1}^{N}\mid\boldsymbol{\theta},T,\left\{z_n\right\}_{n=1}^{N}\right)\propto\sum_{n=1}^{N}\Pr(z_n=0)\log\lambda_0(\mathbf{x}_n) \tag{8.14}$$

$$+\sum_{n=1}^{N}\sum_{n'=1}^{n-1}\Pr(z_n=n')\log\xi(t_n-t_{n'},\mathbf{x}_n-\mathbf{x}_{n'}).$$

Veen and Schoenberg (2008) perform this step by setting the gradient of (8.14) with respect to θ to $\mathbf{0}$ and solving the resulting nonlinear equations. The algorithm continues until $\theta^{(s+1)}-\theta^{(s)}$ is less than some prespecified small number. While the algorithm confers stability and robustness compared to numerical optimization approaches, it does not overcome the $O(N^2)$ time-complexity facing likelihood-based methods: its $\binom{N}{2}$ probabilities $\Pr(z_n=n')$ require $O(N^2)$ floating-point operations. Veen and Schoenberg (2008) successfully apply their algorithm to the analysis of 6,796 earthquakes in Southern California but do not communicate the amount of time required to fit their model.

This strategy solves the aforementioned convergence issues suffered by numerical optimization techniques for finding the maximum likelihood estimator insofar as the EM algorithm provably moves uphill at every iteration, attaining a local maximum upon convergence. Despite the fact that the EM algorithm often results in a consistent estimator (Dempster et al., 1977), the lack of a global maximization guarantee complicates uncertainty quantification. There is no general guarantee of asymptotic normality that one would otherwise use to construct confidence intervals for the EM algorithm's estimator. Within the Bayesian paradigm, one may use the same data augmentation strategy as Veen and Schoenberg (2008) to implement inference algorithms that provide uncertainty quantification with asymptotic exactness guarantees. Example 4 provides one such instance.

The multivariate Hawkes process (MHP) extends the basic HP by introducing marked events $\{t_n,\gamma_n\}_{n=1}^{N}$ for $\gamma_n\in\{1,\ldots,K\}$. MHPs categorize events into one of K classes with K corresponding intensity functions $\{\lambda_k^*(t)\}_{k=1}^{K}$. For any time t, preceding events influence the intensity function in a manner analogous to the basic HP (8.1):

$$\lambda_k^*(t)=\lambda_{0,k}(t)+\sum_{t_n<t}\xi_{\gamma_n,k}(t-t_n). \tag{8.15}$$

Here, $\lambda_{0,k}(t)$ is the background rate of class k, and $\xi_{k',k}(\cdot)$ is the excitation function that describes the contribution of events of class k' to the intensity of class k. The MHP likelihood takes a form analogous to (8.6):

$$L\left(\{t_n,\gamma_n\}_{n=1}^{N} \,\middle|\, \{\lambda_{0,k}(t)\}_{k=1}^{K},\{\xi_{k',k}(t)\}_{k',k=1}^{K},T\right)=e^{-\Lambda(T,K)}\prod_{n=1}^{N}\lambda_{\gamma_n}^{*}(t_n), \tag{8.16}$$

where $\Lambda(T,K)=\int_0^T\sum_{k=1}^{K}\lambda_k^{*}(t)dt$. Its computation has $O(N(N+K))$ time-complexity.

Example 4 (Markov Chain Monte Carlo). *Within a Bayesian hierarchical model formulation, Linderman and Adams (2014) build an MHP with an immigration representation of the likelihood in (8.16):*

$$L\left(\{(t_n,\gamma_n,z_n)\}_{n=1}^{N}\,\middle|\,\{\lambda_{0,k}(t)\}_{k=1}^{K},\{\xi_{k',k}(t)\}_{k',k=1}^{K},T\right)$$
$$=e^{-\Lambda(T,K)}\prod_{n=1}^{N}\lambda_{0,\gamma_n}(t_n)^{\mathcal{I}[z_n=0]}\,\xi_{z_n,\gamma_n}(t_n-t_{z_n})^{\mathcal{I}[z_n\neq0]}. \tag{8.17}$$

Just as in Example 3, $\{z_n\}_{n=1}^{N}$ represent the parents of events $\{t_n\}_{n=1}^{N}$, with $z_n=0$ denoting that event n is an immigrant event. The authors also choose the following parameterization of the baseline intensity and excitation function

$$\lambda_{0,k}(t)=\mu_k+\alpha_k e^{y(t)},\ \ y(t)\sim\mathcal{GP}(0,\kappa(t,t')),\ \ \xi_{k,k'}(t)=A_{k,k'}W_{k,k'}g_{\theta_{k,k'}}(t).$$

A Gaussian process $y(t)$ (Rasmussen and Williams, 2005) and constants μ_k and α_k specify the background rate, while a binary adjacency matrix $\mathbf{A}\in\{0,1\}^{K\times K}$, a matrix of nonnegative weights $\mathbf{W}\in\mathbb{R}_{+}^{K\times K}$, and a logistic-normal density function $g_{\theta_{k',k}}(t)$ with parameters $\{\theta_{k',k}\}_{k',k=1}^{K}$ characterize the excitation function.

Linderman and Adams (2014) develop a novel Gibbs sampler (Geman and Geman, 1984; Gelfand and Smith, 1990) to generate samples from the posterior distribution of the MHP parameters. Gibbs sampling offers a robust procedure for posterior inference by iteratively sampling subsets of parameters from their corresponding conditional posterior distributions. The authors sample the conditional posteriors of **W** *and $\{\theta_{k,k'}\}_{k,k'=1}^{K}$ with specified conjugate priors and $y(t)$ (evaluated on a grid) with elliptical slice sampling (Murray et al., 2010). The authors specify Aldous-Hoover graph priors (Lloyd et al., 2012) on* **A** *to render the elements of* **A** *conditionally independent, but the augmented MHP likelihood* (8.17) *induces dependencies between the rows of* **A**. *Nonetheless, the columns of* **A** *are still independent, enabling the parallelization of Gibbs updates. The authors use collapsed Metropolis-within-Gibbs to sample the columns of* **A** *with the marginalized likelihood* (8.16). *Finally, each parent variable z_n updates from a discrete distribution defined over the set $\{0,1,\ldots,n-1\}$, and all N parent variables are conditionally independent.*

Linderman and Adams (2014) address major computational bottlenecks by pre-specifying the Gaussian process kernel parameters to avoid repeated $O(N^3)$ computations. Additionally, they use parallelization to exploit the conditional independence *between the columns of* $\mathbf{A}$ *and elements of* $\{z_n\}_{n=1}^{N}$ *within their respective blocks. Through parallelization, they reduce the effective time-complexity of updating* $\{z_n\}_{n=1}^{N}$ *from* $O(N^2)$ *to* $O(N)$*, and updating the columns of* $\mathbf{A}$ *from* $O(N^2K)$ *to* $O(N^2)$*. Furthermore, assuming the bounded support of a logistic-normal density reduces the time-complexity of updating* $\mathbf{A}$ *by a significant constant. Finally, parallelized binary reductions summing N contributions to a likelihood reduce* $O(N)$ *to* $O(\log N)$ *time-complexity (cf., Holbrook et al., 2021).*

Linderman and Adams (2014) mention that their parallelized Gibbs sampler achieves 5 to 50 iterations per second for the examples they consider, but they do not report the number of iterations required. The authors implement their algorithm on S&P 100 index dataset collected at 1s intervals during the week from September 28 through October 2, 2009. Events denote times when a stock price changes by ±0.1% of its current price. The dataset has $K = 100$ *processes and* $N = 182{,}037$ *events. They also model gang-related homicides between 1980 and 1995 in Chicago with* $K = 77$ *communities and* $N = 1{,}637$ *homicide events.*

The preceding HPs only allow event occurrences to increase the probability of future events. Nonlinear HPs expand the modeling flexibility of linear HPs with a nonlinear transfer function $f : \mathbb{R} \to \mathbb{R}_+$ that allows one to drop the non-negativity assumption on the excitation function $\xi(\cdot)$ within the conditional intensity function:

$$\lambda^*(t) = f\left(\lambda_0 + \sum_{t_n < t} \xi(t - t_n)\right). \tag{8.18}$$

The inclusion of a nonlinear transfer function allows nonlinear HPs to model inhibitory events and non-additive contributions of past events (Brémaud and Massoulié, 1996). Recent additions to the nonlinear HP literature focus on using deep neural networks to model nonlinear conditional intensities such as (8.18).

Example 5 (Stochastic Gradient Descent). *Mei and Eisner (2017) introduce neural Hawkes processes (NHPs), which they describe as "neurally self-modulating MHPs" that have the ability to model nonlinearities in their conditional intensity functions. For all times* $t > 0$ *and all classes* $k \in \{1, \dots, K\}$*, the class-specific conditional intensity function is*

$$\lambda_k^*(t) = f_k\left(\mathbf{w}_k^\top \mathbf{h}(t)\right),$$

where $f_k(\cdot)$ *is the softplus function,* $\mathbf{w}_k \in \mathbb{R}^D$ *is a vector of weights for class* k*, and* $\mathbf{h}(t) \in (-1,1)^D$ *is a hidden state vector at time* t*. The authors model NHPs using a continuous-time long-short term memory (LSTM) architecture, inspired by the familiar discrete-time LSTM (Hochreiter and Schmidhuber, 1997). In the continuous-time framework, long-term memory* $\mathbf{c}(t)$ *and short-term memory* $\mathbf{h}(t)$ *are*

D-dimensional vectors that update in the continuous interval $(t_n, t_{n+1}]$ *between events* n *and* $n+1$:

$$\mathbf{c}(t) = \bar{\mathbf{c}}_{n+1} + (\mathbf{c}_{n+1} - \bar{\mathbf{c}}_{n+1})e^{(-\delta_{n+1}(t-t_n))}, \quad \mathbf{h}(t) = \mathbf{o}_n \odot (2\sigma(2\mathbf{c}(t)) - 1),$$

where $\sigma(\cdot)$ *is the elementwise sigmoid function and* 0 *is the Hadamard product. Here, long-term memory* $\mathbf{c}(t)$ *exponentially decays at rate* $\boldsymbol{\delta}_{n+1}$ *towards a steady-state value* $\bar{\mathbf{c}}_{n+1}$. *Other LSTM values update instantaneously at time* t_{n+1} *with update rules*

$$\begin{aligned}
\mathbf{i}_{n+1} &\leftarrow \sigma(\mathbf{W}_i\mathbf{k}_n + \mathbf{U}_i\mathbf{h}(t_n) + \mathbf{d}_i) & \bar{\mathbf{i}}_{n+1} &\leftarrow \sigma(\mathbf{W}_{\bar{i}}\mathbf{k}_n + \mathbf{U}_{\bar{i}}\mathbf{h}(t_n) + \mathbf{d}_{\bar{i}}) \\
\mathbf{f}_{n+1} &\leftarrow \sigma(\mathbf{W}_f\mathbf{k}_n + \mathbf{U}_f\mathbf{h}(t_n) + \mathbf{d}_f) & \bar{\mathbf{f}}_{n+1} &\leftarrow \sigma(\mathbf{W}_{\bar{f}}\mathbf{k}_n + \mathbf{U}_{\bar{f}}\mathbf{h}(t_n) + \mathbf{d}_{\bar{f}}) \\
\mathbf{z}_{n+1} &\leftarrow 2\sigma(\mathbf{W}_z\mathbf{k}_n + \mathbf{U}_z\mathbf{h}(t_n) + \mathbf{d}_z) - 1 & \mathbf{c}_{n+1} &\leftarrow \mathbf{f}_{n+1} \odot \mathbf{c}(t_n) + \mathbf{i}_{n+1} \odot \mathbf{z}_{n+1} \\
\mathbf{o}_{n+1} &\leftarrow \sigma(\mathbf{W}_o\mathbf{k}_n + \mathbf{U}_o\mathbf{h}(t_n) + \mathbf{d}_o) & \bar{\mathbf{c}}_{n+1} &\leftarrow \bar{\mathbf{f}}_{n+1} \odot \bar{\mathbf{c}}_n + \bar{\mathbf{i}}_{n+1} \odot \mathbf{z}_{n+1} \\
\delta_{n+1} &\leftarrow \zeta(\mathbf{W}_d\mathbf{k}_n + \mathbf{U}_d\mathbf{h}(t_n) + \mathbf{d}_d).
\end{aligned}$$

for nonlinear function $\zeta : \mathbb{R} \to \mathbb{R}_+$. *The vector* $\boldsymbol{k}_n \in \{0,1\}^K$ *is the one-hot encoding of the class of event* n, *and* $\{\boldsymbol{W}_\ell \in \mathbb{R}^{D\times K}, \boldsymbol{U}_\ell \in \mathbb{R}^{D\times D}, \boldsymbol{d}_\ell \in \mathbb{R}^D\}$ *are LSTM parameters for* $\ell \in \{i, f, z, o, d, \bar{i}, \bar{f}\}$.

Mei and Eisner (2017) use stochastic gradient descent (SGD) (Bottou, 2010) to fit the LSTM parameters by minimizing the negative log-likelihood (cross-entropy loss)

$$-\ell\left(\{t_n, \gamma_n\}_{n=1}^N \,|\{\lambda_{0,k}(t)\}_{k=1}^K, \{\xi_{k',k}(t)\}_{k',k=1}^K, T\right) = \Lambda(T,K) - \sum_{n=1}^N \log \lambda^*_{\gamma_n}(t_n)$$

arising from (8.16). In each iteration of SGD, the authors use Monte Carlo simulations to estimate the gradient of the cumulative intensity, $\nabla\Lambda(T,K)$:

$$\nabla\hat{\Lambda}(T,K) = \frac{T}{M}\sum_{m=1}^{M}\sum_{k=1}^{K}\nabla\lambda^*_k(t_m), \quad t_1, t_2, \ldots, t_M \overset{iid}{\sim} Uniform(0,T). \tag{8.19}$$

The estimate $\nabla\hat{\Lambda}(T,K)$ *is an unbiased estimator of* $\nabla\Lambda(T,K)$ *since it is the sum of MK true gradients. The computational bottleneck of SGD lies in the computation of* $\hat{\Lambda}$ *with a time-complexity of* $O(MND(D+K))$, *where* D *is the dimensionality of the LSTM memory. Computing* $\nabla\lambda^*_k(t_m)$ *in (8.19) requires forward-pass through* $O(N)$ *LSTM cells, each of which performs two matrix multiplication steps with* $O(D^2)$ *and* $O(DK)$ *time-complexity. Mei and Eisner (2017) fit their model with Theano, a now-defunct Python library built to enable massively parallel GPU-based* *acceleration of key backpropagation computations. The authors fit their model on a Twitter dataset (Zhao et al., 2015)* $(K = 3, N = 166{,}076)$ *and the MemeTrack dataset (Leskovec and Krevl, 2014)* $(K = 5{,}000, N = 1{,}500{,}000)$*, but they do not mention the time required to do so.*

8.3 Analysis of American Gun Violence Data

In this section, we use data from the American gun violence epidemic to demonstrate how high-performance computing (HPC) can help overcome the HP's usual quadratic time-complexity to enable HP-based analyses of large-scale public health data. We model spatiotemporal data $\{(\mathbf{x}_n, t_n)\}_{n=1}^{N}$ similar to that of Example 3 using an StHP with conditional intensity function

$$\lambda^*(t, x) = \frac{\lambda_0}{|\mathcal{X}|} \sum_{n=1}^{N} \mathcal{I}_{[t \neq t_n]} \phi_1(t \mid t_n, \tau_t) + \frac{\xi_0}{\omega_t} \sum_{t_n < t} e^{-(t-t_n)/\omega_t} \phi_2(\mathbf{x} \mid \mathbf{x}_n, \omega_x), \quad (8.20)$$

with $\mathcal{I}$ the indicator function, $|\mathcal{X}|$ the area of the spatial domain, ϕ_1 the one-dimensional Gaussian density, ϕ_2 the two-dimensional spherical Gaussian density, and $\lambda_0, \tau_t, \omega_x, \omega_t$ and ξ_0 positive parameters. Here, τ_t is the background temporal lengthscale, ω_t is the triggering kernel's temporal lengthscale, and ω_x is the triggering kernel's spatial lengthscale. We use the same prior specification for these parameters as that of Ko et al. (2024), but note that priors appear to have little influence in the big data context we consider. Namely, we model 412,376 acts of gun violence recorded in the US between 2014 and 2024 (Figure 8.1).

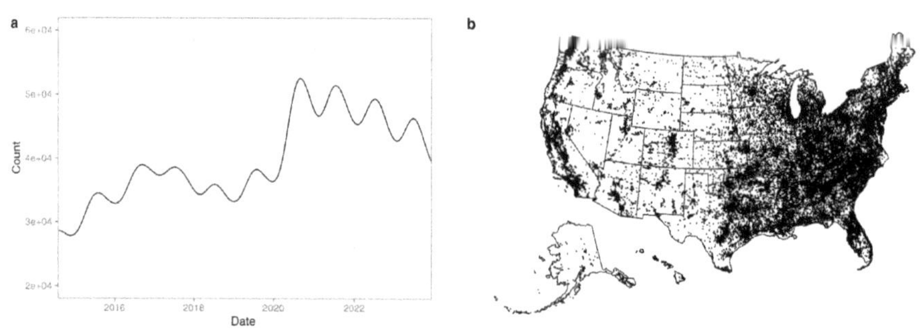

FIGURE 8.1
Temporal (a) and Spatial (b) Distributions of 412,376 American Gun Violence Events Recorded from 2014 to 2024 and Provided by the Gun Violence Archive. We use the high-performance computational inference strategy of Ko et al. (2024) to perform Bayesian inference and analyze this data in only 10 hours, despite the task requiring more than $O(16 \times 10^{10})$ floating-point operations per iteration of Markov chain Monte Carlo.

We fit our StHP model to this large-scale spatiotemporal data using the computational strategy for Bayesian inference advanced in Ko et al. (2024). As is the case for many of the likelihood-based methods discussed in Section 8.2, the likelihood associated with (8.20) scales $O\left(N^2\right)$ in time-complexity. Following Ko et al. (2024), we use a multi-GPU extension of the single-GPU likelihood implementation of Holbrook et al. (2021). Figure 8.2 presents the relative performance of likelihood implementations using different numbers of GPUs; notably, the single-GPU likelihood implementation is already hundreds of times faster than an efficient C++ likelihood implementation and thousands of times faster than an analogous implementation in R (Holbrook et al., 2021). We use four GPU nodes, each containing eight AMD Radeon Instinct MI50 GPUs, to generate 31 adaptive MH chains, with two GPUs assigned to each of the 16 chains run simultaneously. Individual chains require 4.11 hours on average to complete, and the entire batch of chains less than 10 hours. Across model parameters, we see a maximum $\hat{R}$ less than 1.003, minimum bulk- and tail-ESSs above 10,000 and 15,000, respectively, and minimum single-chain ESS above 300 (Stan Development Team, 2024; Plummer et al., 2006).

Table 8.1 presents the empirical posterior medians and 95% credible intervals $\left(CrI\right)$ for each of the model parameters. We are particularly interested in the posterior distributions of ξ_0,ω_x and ω_t. Because the temporal and spatial kernels used within the triggering kernel are normalized densities, the self-excitatory weight ξ_0 corresponds exactly to the branching number β of (8.2). Thus, we conclude that the observed gun violence process is explosive: we expect each event to birth an additional 1.15 (95% CrI: 1.15, 1.16) events. The posterior median for the expected spatial distance between an event and its

TABLE 8.1

Posterior Medians and 95% Credible Intervals for Parameters of a Spatiotemporal Hawkes Process Model with Conditional Intensity (8.20) Applied to American Gun Violence Data Visualized in Figure 8.1

Rate Component	Parameter	Posterior Median (95% CrI)
Background	Weight λ_0	0.0231 (0.0227, 0.0237)
	Temporal lengthscale τ_t (wks)	2.62 (2.36, 2.93)
Self-excitatory	Weight ξ_0	1.15 (1.15, 1.16)
	Spatial lengthscale ω_x (deg*)	0.0205 (0.0204, 0.0206)
	Temporal lengthscale ω_t (wks)	75.8 (75.1, 76.5)

Note: The unit of measurement for the spatial lengthscale parameter ω_x is degrees (latitude/longitude), for which the posterior median of 0.0205* converts to, e.g., 1.35 miles in Miami, FL, and 1.19 miles in Seattle, WA.

child event, conveyed by ω_x, is 0.0205 degrees. When translated to miles, the number varies depending on the latitude. For example, 0.0205 degrees translates to 1.35 miles in Miami, FL, and 1.19 miles in Seattle, WA. Finally, the posterior median for the expected time lapse between an event and its child event, conveyed by ω_t, is 75.8 weeks ($95\% CrI$: 75.1, 76.5) or almost 1.5 years. This number is orders-of-magnitude larger than the analogous number cited in Holbrook et al. (2021), but that analysis involves gunshots recorded by AGLS and not the acts of gun violence considered here.

8.4 Discussion

We have shown how to scale Bayesian inference to big data for a relatively simple StHP, but there is much work to be done in the advancement of computational tools for fitting HPs at scale. In particular, there are major tensions between making HP models scalable and making them robust, flexible, or expressive. Furthermore, incorporating interesting structures underlying data, such as networks or multiple sources, often brings added computational complexity, thereby making fitting HP models adapted to these structures all the more difficult (e.g., Example 4). We also see major tensions between scalability and user-friendliness: the large-scale analysis of gun violence in Section 8.3 relies on hand-coded GPU kernels written in a language (OpenCL) that extends the C programming language. NHPs such as that of Example 5 are flexible and programmable using widely used deep learning platforms such as PyTorch or TensorFlow, but their reliance on stochastic gradient descent for scalability means a lack of rigorous model identification and uncertainty quantification. Nonetheless, we see NHPs and other deep generative models as the dominant modeling paradigm for event forecasting moving forward, with non-neural modeling reserved for interpretable and rigorous scientific inference that sheds light on natural mechanisms and, perhaps, informs policy.

Figure 8.2 Absolute and Relative Speedups for Spatiotemporal Hawkes Process Likelihood (8.11) with Conditional Intensity (8.20) Using Different Numbers of GPUs. We make the right plot using 412,376 data points, but trends hold for as many as 1,000,000. Notably, the single-GPU implementation is already over 200-times faster than an efficient single-core C++ implementation. Multi-GPU likelihood evaluations accelerate individual Metropolis-Hastings chains within our many-GPU, many-chain Bayesian inference strategy.

The ability to fit and compare a variety of competing HP models is crucial for meaningful scientific inference and currently lacking in big data contexts. In Section 8.3, we fit the StHP with rate function (8.20) to data consisting of

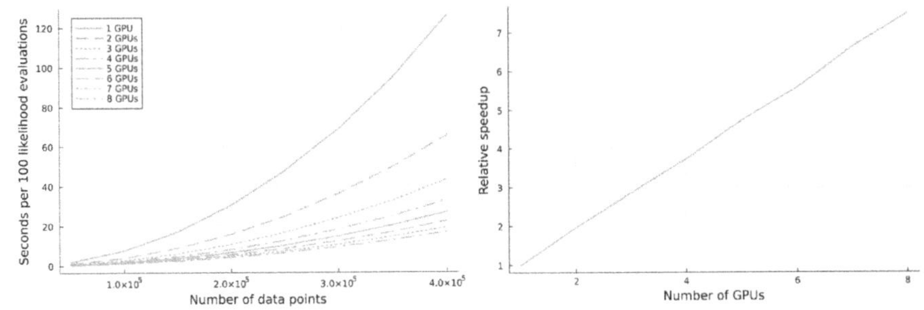

FIGURE 8.2
Absolute and relative speedups for spatiotemporal Hawkes process likelihood (11) with conditional intensity (20) using different numbers of GPUs. We make the right plot using 412,376 data points, but trends hold for as many as 1,000,000. Notably, the single- GPU implementation is already over 200-times faster than an efficient single-core C++ implementation. Multi-GPU likelihood evaluations accelerate individual Metropolis-Hastings chains within our many-GPU, many-chain Bayesian inference strategy.

over 400k gun violence events, inferring scientifically interpretable parameters that shed light on the expanding nature of gun violence in the US (namely, we estimate that an average of 1.15 acts of violence arise from any individual act of violence) and the scales of its local spatial contagion. While we remain enthusiastic about the ability of increasingly widespread high-performance computing technology to scale HP to big data, we cannot fully trust our inference, not having compared our model to other competitors. Notably, we find the extremely narrow posterior credible intervals of Table 8.1 particularly distressing. Probably, the use of Bayesian nonparametric priors to define HP triggering kernels may help address this dearth of posterior uncertainty, but one needs to thoughtfully engage parallel computing resources for algorithmic developments that go hand in hand with model extensions. The same statement applies for more prosaic modeling decisions such as that between exponential, power-law, and Weibull temporal triggering kernels.

Many criteria for model comparison exist, but we believe that a model's forecasting accuracy is the only relevant criterion for big point process data. This is because datasets consisting of many temporal events often become large when they span large swaths of time over which stationarity assumptions fail. Ideally, one would be able to train a candidate model on a nested sequence of data subsets, measuring different forecasting accuracy on subsequent data subsamples of varying time lengths. One might divide the gun violence data analyzed in Section 8.3 into nested subsets increasing at 1-month intervals, using HP model instances trained on each subset to generate probabilistic forecasts, say, 4 weeks into the future. With 10 years of data, one would need to train each candidate model on roughly 120 subsets and use trained models to simulate large numbers of events. It is for these reasons we claim that proper scientific inference using HP applied to large sequences of events requires methods and computational tools for simulating HP sample paths at scale, a subject we see nowhere in the literature.

Acknowledgments

AJH is supported by grants NIH K25 AI153816, NSF DMS 2152774, and NSF DMS 2236854. We gratefully acknowledge generous support of Advanced Micro Devices, Inc., including the donation of parallel computing resources used for this research. We also thank the Gun Violence Archive for sharing its data.

References

Bottou, L. (2010). Large-scale machine learning with stochastic gradient descent. In *Proceedings of COMPSTAT'2010: 19th International Conference on Computational Statistics Paris France,* August 22–27, 2010 Keynote, Invited and Contributed Papers, pp. 177–186. Springer.

Brantingham, P. J., J. Carter, J. MacDonald, C. Melde, and G. Mohler (2021). Is the recent surge in violence in american cities due to contagion? *Journal of Criminal Justice* 76, 101848.

Brémaud, P. and L. Massoulié (1996). Stability of nonlinear Hawkes processes. *The Annals of Probability* 24(3), 1563–1588.

Chavez-Demoulin, V. and J. McGill (2012). High-frequency financial data modeling using hawkes processes. *Journal of Banking & Finance* 36(12), 3415–3426.

Choi, E., N. Du, R. Chen, L. Song, and J. Sun (2015). Constructing disease network and temporal progression model via context sensitive Hawkes process. In 2015 *IEEE International Conference on Data Mining,* pp. 721–726. IEEE.

Da Fonseca, J. and R. Zaatour (2014). Hawkes process: Fast calibration, application to trade clustering, and diffusive limit. *Journal of Futures Markets* 34(6), 548–579.

Daley, D. J. and D. Vere-Jones (2003). Conditional intensities and likelihoods. In *An Introduction to the Theory of Point Processes: Volume I: Elementary Theory and Methods,* pp. 211–287.

Daw, A. and J. Pender (2023). Matrix calculations for moments of Markov processes. *Advances in Applied Probability* 55(1), 126–150.

Dempster, A. P., N. M. Laird, and D. B. Rubin (1977). Maximum likelihood from incomplete data via the EM algorithm. *Journal of the Royal Statistical Society: Series B (Methodological)* 39(1), 1–22.

Dong, Z., X. Cheng, and Y. Xie (2023). Spatio-temporal point processes with deep non-stationary kernels. In *The Eleventh International Conference on Learning Representations,* Kigali, Rwanda.

Embrechts, P., T. Liniger, and L. Lin (2011). Multivariate hawkes processes: An application to financial data. *Journal of Applied Probability* 48(A), 367–378.
Fagan, J., D. L. Wilkinson, and G. Davies (2007). Social contagion of violence. In D. J. Flannery, A. T. Vazsonyi, and I. D. Waldman (Eds.). *Cambridge Handbooks in Psychology*. Cambridge University Press, pp. 688–724.
Gelfand, A. and A. Smith (1990). Sampling-based approaches to calculating marginal densities. *Journal of the American Statistical Association* 85, 398–409.
Geman, S. and D. Geman (1984). Stochastic relaxation, gibbs distributions, and the bayesian restoration of images. *IEEE Transactions on Pattern Analysis and Machine Intelli-gence PAMI* 6(6), 721–741.
Hall, A. R. (2005). *Generalized Method of Moments* (advanced texts in econometrics series). Oxford University Press.
Hardiman, S. J., N. Bercot, and J.-P. Bouchaud (2013). Critical reflexivity in financial markets: A hawkes process analysis. *The European Physical Journal B* 86(10), 442.
Hastings, W. K. (1970). Monte carlo sampling methods using markov chains and their applications. *Biometrika* 57, 97–109.
Hawkes, A. G. (1971a). Point spectra of some mutually exciting point processes. *Journal of the Royal Statistical Society: Series B (Methodological)* 33(3), 438–443.
Hawkes, A. G. (1971b). Spectra of some self-exciting and mutually exciting point processes. *Biometrika* 58(1), 83–90.
Hawkes, A. (1972). Spectra of some mutually exciting point processes with associated variables. In Peter A. W. Lewis(Ed.), *Stochastic Point Processes*, pp. 261–271.
Hawkes, A. and Adamopoulos, L. (1973). Cluster models for earthquakes-regional comparisons. *Bulletin of the International Statistical Institute* 45(3), 454–461.
Hawkes, A. G. (2018). Hawkes processes and their applications to finance: A review. *Quantitative Finance* 18(2), 193–198.
Hawkes, A. G. and D. Oakes (1974). A cluster process representation of a self-exciting process. *Journal of Applied Probability* 11(3), 493–503.
Hochreiter, S. and J. Schmidhuber (1997). Long short-term memory. *Neural Computation* 9(8), 1735–1780.
Holbrook, A. J., X. Ji, and M. A. Suchard (2022a). Bayesian mitigation of spatial coarsening for a hawkes model applied to gunfire, wildfire and viral contagion. *The Annals of Applied Statistics* 16(1), 573–595.
Holbrook, A. J., X. Ji, and M. A. Suchard (2022b). From viral evolution to spatial contagion: A biologically modulated hawkes model. *Bioinformatics* 38(7), 1846–1856.
Holbrook, A. J., P. Lemey, G. Baele, S. Dellicour, D. Brockmann, RambautA. , and M. A. Suchard (2021a). Massive parallelization boosts big bayesian multidimensional scaling. *Journal of Computational and Graphical Statistics* 30(1), 11–24.
Holbrook, A. J., C. E. Loeffler, S. R. Flaxman, and M. A. Suchard (2021b). Scalable bayesian inference for self-excitatory stochastic processes applied to big american gunfire data. *Statistics and Computing* 31(1), 1–15.
Kelly, J. D., J. Park, R. J. Harrigan, N. A. Hoff, S. D. Lee, R. Wannier, B. Selo, M. Mossoko, B. Njoloko, E. Okitolonda-Wemakoy, et al. (2019). Real-time predictions of the 2018–2019 ebola virus disease outbreak in the democratic republic of the congo using hawkes point process models. *Epidemics* 28, 100354.
Kim, H. (2011). Spatio-temporal point process models for the spread of avian influenza virus (H5N1). Ph.D. thesis, UC Berkeley.
Ko, S., M. A. Suchard, and A. J. Holbrook (2024). Scaling Hawkes processes to one million COVID-19 cases. arXiv preprint arXiv:2407.11349.

Laub, P. J., Y. Lee, and T. Taimre (2021). *The Elements of Hawkes Processes*. Springer.

Laub, P. J., T. Taimre, and P. K. Pollett (2015). Hawkes processes. arXiv preprint arXiv:1507.02822.

Leskovec, J. and A. Krevl (2014, June). SNAP datasets: Stanford large network dataset collection. http://snap.stanford.edu/data.

Levenberg, K. (1944). A method for the solution of certain non-linear problems in least squares. *Quarterly of Applied Mathematics* 2(2), 164–168.

Linderman, S. and R. Adams (2014). Discovering latent network structure in point process data. In *International Conference on Machine Learning*, pp. 1413–1421. PMLR.

Linderman, S. W., Y. Wang, and D. M. Blei (2017). Bayesian inference for latent hawkes processes. In *Advances in Approximate Bayesian Inference Workshop*, 31st Conference on Neural Information Processing Systems (NIPS 2017) .

Lloyd, J., P. Orbanz, Z. Ghahramani, and D. M. Roy (2012). Random function priors for ex-changeable arrays with applications to graphs and relational data. In F. Pereira, C. Burges, L. Bottou, and K. Weinberger (Eds.), *Advances in Neural Information Processing Systems*, Volume 25. Curran Associates, Inc.

Loeffler, C. and S. Flaxman (2018). Is gun violence contagious? A spatiotemporal test. *Journal of Quantitative Criminology* 34(4), 999–1017.

Loftin, C. (1986). Assaultive violence as a contagious social process. *Bulletin of the New York Academy of Medicine* 62(5), 550.

Marquardt, D. W. (1963). An algorithm for least-squares estimation of nonlinear parameters. *Journal of the Society for Industrial and Applied Mathematics* 11(2), 431–441.

Mei, H. and J. M. Eisner (2017). The neural hawkes process: A neurally self-modulating multivariate point process. In I. Guyon, U. Von Luxburg, S. Bengio, H. Wallach, R. Fergus, VishwanathanS. , and R. Garnett (Eds.), *Advances in Neural Information Processing Systems*, pp. 6754–6764.

Metropolis, N., A. W. Rosenbluth, M. N. Rosenbluth, A. H. Teller, and E. Teller (1953). Equation of state calculations by fast computing machines. *The Journal of Chemical Physics* 21(6), 1087–1092.

Meyer, S., L. Held, et al. (2014). Power-law models for infectious disease spread. *The Annals of Applied Statistics* 8(3), 1612–1639.

Mohler, G. (2014). Marked point process hotspot maps for homicide and gun crime prediction in chicago. *International Journal of Forecasting* 30(3), 491–497.

Murray, I., R. P. Adams, and D. J. MacKay (2010). Elliptical slice sampling. *Journal of Machine Learning Research: Workshop and Conference Proceedings* 9, 541–548.

Ogata, Y. (1988). Statistical models for earthquake occurrences and residual analysis for point processes. *Journal of the American Statistical Association* 83(401), 9–27.

Ozaki, T. (1979). Maximum likelihood estimation of hawkes' self-exciting point processes. *Annals of the Institute of Statistical Mathematics* 31(1), 145–155.

Park, J., F. P. Schoenberg, A. L. Bertozzi, and P. J. Brantingham (2021). Investigating clustering and violence interruption in gang-related violent crime data using spatial–temporal point processes with covariates. *Journal of the American Statistical Association* 116(536), 1674–1687.

Plummer, M., N. Best, K. Cowles, and K. Vines (2006). Coda: Convergence diagnosis and output analysis for mcmc. *R News* 6(1), 7–11.

Rasmussen, C. E. and C. K. I. Williams (2005, 11). *Gaussian Processes for Machine Learning*. The MIT Press.

Reinhart, A. (2018). A review of self-exciting spatio-temporal point processes and their applications. *Statistical Science* 33(3), 299–318.

Rizoiu, M.-A., S. Mishra, Q. Kong, M. Carman, and L. Xie (2018). Sir-Hawkes: Linking epidemic models and Hawkes processes to model diffusions in finite populations. In *Proceedings of the 2018 World Wide Web Conference on World Wide Web*, pp. 419–428. International World Wide Web Conferences Steering Committee.

Rubin, I. (1972). Regular point processes and their detection. *IEEE Transactions on Information Theory* 18(5), 547–557.

Sha, H., M. Al Hasan, J. Carter, and G. Mohler (2020). Interpretable hawkes process spatial crime forecasting with tv-regularization. In *2020 IEEE International Conference on Big Data (Big Data)*, pp. 3228–3236. IEEE.

Stan Development Team (2024). RStan: The R interface to Stan. R package version 2.32.5.

Taylor, R. M., M. A. Simon, and D. M. Patel (2013). *Contagion of Violence: Workshop Summary*. National Academies Press.

Truccolo, W. (2016). From point process observations to collective neural dynamics: Non-linear hawkes process glms, low-dimensional dynamics and coarse graining. *Journal of Physiology-Paris* 110(4), 336–347.

Veen, A. and F. P. Schoenberg (2008). Estimation of space-time branching process models in seismology using an em–type algorithm. *Journal of the American Statistical Association* 103(482), 614–624.

Yang, S.-H. and H. Zha (2013). Mixture of mutually exciting processes for viral diffusion. In *Proceedings of the 30th International Conference on Machine Learning, PMLR 28 (2), 1–9.*

Zhao, Q., M. A. Erdogdu, H. Y. He, A. Rajaraman, and J. Leskovec (2015). Seismic: A self-exciting point process model for predicting tweet popularity. In *Proceedings of the 21th ACM SIGKDD International Conference on Knowledge Discovery and Data Mining, KDD '15*, New York, NY, USA, pp. 1513–1522. Association for Computing Machinery.

Zhu, S. and Y. Xie (2022). Spatiotemporal-textual point processes for crime linkage detection. *The Annals of Applied Statistics* 16(2), 1151–1170.

Zhuang, J., Y. Ogata, and D. Vere-Jones (2004). Analyzing earthquake clustering features by using stochastic reconstruction. *Journal of Geophysical Research: Solid Earth* 109 (B5), B05301.

9

Atlanta Gun Violence Modeling via Nonstationary Spatio-Temporal Point Processes

Zheng Dong and Yao Xie

Analysis of gun violence in the United States has utilized various models based on spatio-temporal point processes. Previous studies have identified a contagion effect in gun violence, characterized by bursts of diffusion across urban environments, which can be effectively represented using the self-excitatory spatio-temporal Hawkes process. The Hawkes process and its variants have been successful in modeling self-excitatory events, including earthquakes, disease outbreaks, financial market movements, neural activity, and the viral spread of memes on social networks. However, existing Hawkes models applied to gun violence often rely on simplistic stationary kernels, which fail to account for the complex, non-homogeneous spread of influence and impact over space and time. To address this limitation, we adopt a non-stationary spatio-temporal point process model that incorporates a neural network-based kernel to better represent the varied correlations among events of gun violence. Our study analyzes a comprehensive dataset of approximately 16,000 gunshot events in the Atlanta metropolitan area from 2021 to 2023. The cornerstone of our approach is the innovative non-stationary kernel, designed to enhance the model's expressiveness while preserving its interpretability. This approach not only demonstrates strong predictive performance but also provides insights into the spatio-temporal dynamics of gun violence and its propagation within urban settings.

9.1 Introduction

Gun violence remains a critical issue in many US cities, influenced by a complex interplay of social and economic factors (Braga et al., 2010). Understanding the patterns and determinants of gun violence is essential

DOI: 10.1201/9781003430773-10

for local policymakers to develop and implement effective intervention strategies. Previous research on gun-related incidents (Fagan et al., 2007; Green et al., 2017; Loeffler and Flaxman, 2018; Papachristos, 2009; Rosenfeld et al., 1999; Short et al., 2014) has consistently identified a contagious pattern of gun violence, suggesting that these events are not isolated but are dependent through underlying social and spatial factors. Although most gun-related incidents do not exhibit a contagion effect, a subset of events show patterns consistent with contagion (Loeffler and Flaxman, 2018), highlighting an important mechanism through which gun violence propagates. These patterns present a predictive potential that can be leveraged to mitigate future risks of gun violence. Additionally, the availability of rich, real-time data sources in metropolitan areas, such as 911 call records, provides a valuable opportunity for analyzing the underlying dynamics of gun-related events and forecasting the risk of gun violence in urban settings.

However, modeling and predicting gun violence continue to present significant challenges. A primary obstacle lies in the vast volume and complexity of the data involved. For instance, our dataset comprises nearly 16,000 reported gunshot events in the city of Atlanta, creating a substantial computational burden and necessitating sophisticated analytical methods. Furthermore, the data is multi-modal, as it integrates direct reports from 911 calls with detailed demographic and socioeconomic census data. Previous studies have addressed the spatio-temporal clustering of gun violence (Loeffler and Flaxman, 2018; Taylor et al., 2013) and other types of crime (Cohen and Tita, 1999; Mohler et al., 2011; Morenoff et al., 2001; Johnson, 2008; Short et al., 2010), but they often employ simpler models that cannot fully capture the heterogeneous spatio-temporal dynamics of violence spread across different urban contexts. This gap underscores the need for a more tailored framework capable of effectively characterizing these complex event patterns.

In this chapter, we present a point process model designed to address the challenges of modeling and capturing the complex spatio-temporal dynamics underlying gunshot events, and showcase using gunshots data in Atlanta in the United States. Drawing inspiration from previous non-stationary point process models for earthquakes and pandemics (Dong et al., 2023c; Zhu et al., 2022b), our approach incorporates a non-stationary influence kernel to represent the intricate contagious effects of gunshot events. Moreover, our model extends beyond existing frameworks by integrating a heterogeneous background intensity, which accounts for the inherent variability in gunshot events influenced by diverse census factors across different geographic regions. Through validation using real-world data, we demonstrate the effectiveness and predictive power of our model compared with the state of the art. The proposed model can be used to understand gun violence propagation and also enhances our capacity to predict future hotspots and trends, thereby providing actionable insights for law enforcement and public policymakers to prevent urban gun violence.

The rest of the chapter is organized as follows. In the remainder of this section, we continue to review related works. Section 9.2 introduces the gunshot data collected by the Atlanta Police Department and the public census data used in our study. Section 9.3 presents the point process model; Section 9.4 explains modeling fitting strategies. Section 9.5 presents numerical results on the real datasets to demonstrate the effectiveness and interpretability of our model. Finally, Section 9.6 concludes the chapter with discussions.

9.1.1 Related Works

The contagious effect of crime has been previously studied since several early works (Fagan et al., 2007; Loftin, 1986; Roth and Reiss Jr, 1993), where researchers modeled gun-related homicides as a point process with contagion characteristics. Crime contagion refers to the phenomenon where the occurrence of violence increases the likelihood of subsequent violence. Notably, most researchers have focused on this contagion at the event level, examining its effects over shorter time scales. They emphasize the contribution of individual events to the heightened crime risk in the following days and in nearby neighborhoods (Skogan et al., 2008; Webster et al., 2013), which is part of the broader spatio-temporal study of crime properties.

Spatio-temporal clustering patterns of crime events have been documented across various crime types, including violent crime (Braga et al., 2010; Cohen and Tita, 1999; Morenoff et al., 2001; Ratcliffe and Rengert, 2008) and property crime (Johnson, 2008; Mohler et al., 2011; Short et al., 2010). Early work by Messner et al. (1999) identified a non-random clustering pattern of homicides at the county level, while Ratcliffe and Rengert (2008) observed similar non-random spatio-temporal clustering at the block level. In contrast to these studies, our research focuses on analyzing violent crime patterns at the event level, without spatial aggregation of crime data. A recent study (Loeffler and Flaxman, 2018) highlights the difference between the clustering of gun violence with and without diffusion/contagion effects, demonstrating the existence of such a spatio-temporal diffusion pattern of gun violence in a metropolitan area.

The application of self-exciting point processes in crime modeling, originally motivated by their use in modeling earthquake occurrences in seismology (Ogata, 1988), has been extensively explored to characterize the spatiotemporal clustering patterns of crime events (Lewis et al., 2012; Mohler, 2013; Mohler et al., 2011; Reinhart, 2018; Zhu and Xie, 2022; Zhuang and Mateu, 2019). Early attempts demonstrated the effectiveness of point processes in modeling crime using residential burglary data in Los Angeles (Mohler et al., 2011) and civilian death reports in Iraq (Lewis et al., 2012). Subsequent approaches (Mohler, 2014; Reinhart and Greenhouse, 2018; Zhu and Xie, 2022) improved point process models for crime by incorporating the events' type, location, and textual information to capture complex crime patterns in various modeling tasks. Our approach extends the commonly used stationary influence kernel to a non-stationary one with a neural network representation, enhancing the model's ability to capture the spatial heterogeneity observed in the occurrences of gunshot events. The effectiveness of our non-stationary kernel is both theoretically and empirically supported by previous studies on deep non-stationary influence kernels (Dong et al., 2023a, b, c; Zhu et al., 2022b) and the numerical results presented in this study.

An increasing level of gun violence has been observed in many US cities, drawing significant attention from the criminology community. Previous studies (Fagan et al., 2007; Loeffler and Flaxman, 2018; Taylor et al., 2013) have suggested the presence of gun violence contagion, which can be driven by various factors, such as the rapid diffusion of human behaviors (De Tarde, 1903) or the structure of social networks (Green et al., 2017; Papachristos, 2009; Rosenfeld et al., 1999; Short et al., 2014). Another line of research highlights a strong relationship between gun violence and the characteristics of local communities, including socioeconomic factors (Braga et al., 2010; Kravitz-Wirtz et al., 2022; Morenoff et al., 2001; Papachristos, 2009) and physical infrastructure (Xu and Griffiths, 2017). The goal of this work is to model the dynamics of gun violence in the City of Atlanta. This is achieved by considering both the contagion of gunshot events and the influence of various exogenous spatial factors across communities on the risk of gun violence.

9.2 Atlanta Gunshot Data

The gunshot event dataset used in this study was collected by the Atlanta Police Department (APD) in the city of Atlanta, the capital and most populous city in Georgia, and the core of the sixth-largest US metropolitan area. To foster a manageable and supportive living environment, the city is divided into 242 distinct neighborhoods, including notable areas such as Midtown, Downtown, and Buckhead. In this study, we base our data analysis and model development on this neighborhood division, enabling us to understand the influence of local factors on gunshot patterns and to support the development of tailored strategies for enhancing community safety.

The gunshot dataset comprises a total of 15,940 reported gunshot incidents from May 11, 2021, to May 10, 2023. Each incident's time (accurate to the second) and geographic location (measured in longitude and latitude) are recorded. Figure 9.1 illustrates the daily number of gunshot events over this two-year period. Although the number of gunshot events remains relatively stable over time, there is a noticeable surge in intensity around major holidays. Figure 9.2 visualizes the spatial distribution of all reported gunshot events within the Atlanta city area, both at the event level and neighborhood level. The spatial heterogeneity of gunshot events is evident, with the center (*e.g.*, Downtown) and southern parts of Atlanta experiencing more reported

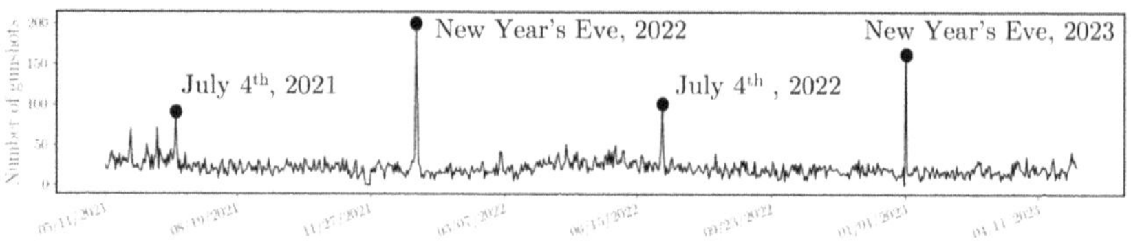

FIGURE 9.1
Daily Number of Gunshot Events from May 11, 2021, to May 10, 2023, with Four Peaks of Gunshot Events Highlighted in the Line Chart.

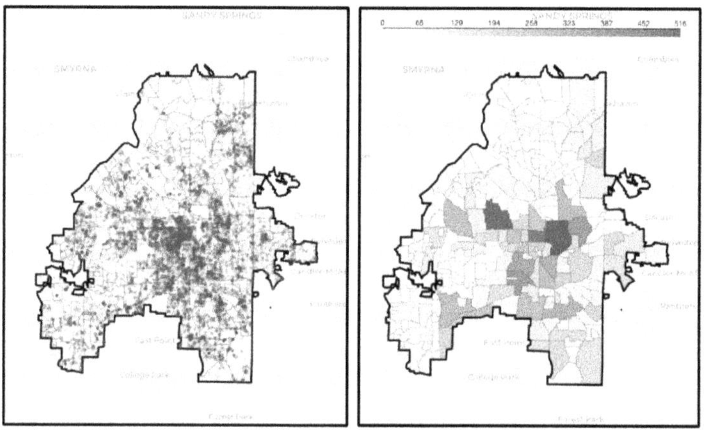

(a) All gunshot events (b) Aggregated by neighborhood

FIGURE 9.2
Distribution of All Gunshot Events Reported in the Atlanta City Area from May 11, 2021, to May 10, 2023. The gray lines indicate the boundaries of different neighborhoods. (a) Event-level distribution: Each dot represents a reported gunshot event, with darker areas indicating overlapping dots. (b) Neighborhood-level distribution: The number of gunshot events in each neighborhood is visualized, with darker shading indicating higher numbers of gunshot events.

incidents than the northern part of the city (*e.g.*, Buckhead). This non-stationarity in both temporal and spatial dimensions highlights the need for a non-stationary kernel and location-dependent background intensity in the model to account for the potential effects of significant spatial covariates.

9.2.1 Why Are Non-stationary Spatio-temporal Models Necessary for Our Data?

In our scenario, point process models with a stationary influence kernel (*e.g.*, Epidemic-Type Aftershock Sequence model (ETAS) (Ogata, 1988)) may lack the necessary flexibility. A stationary influence kernel is shift-invariant, meaning its value depends only on the difference between two time points or locations. Consequently, it cannot capture non-isotropic spatial influences or other complex spatial dependence structures. For example, gunshot events

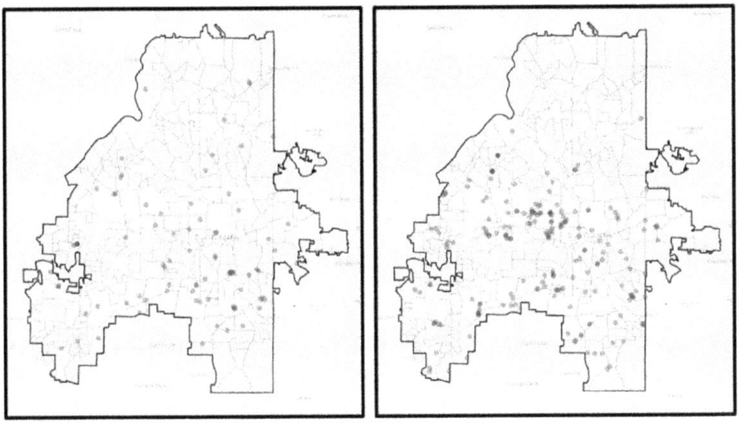

(a) July 4th, 2021 (b) December 31st, 2021

FIGURE 9.3
Snapshots of the Spatial Distributions of Gunshot Events on Independence Day and New Year's Eve in 2021, Respectively.

in Atlanta are heterogeneously distributed across different neighborhoods with a complex structure, as shown in Figure 9.2. Moreover, the contagious effect of gunshot events is non-stationary and significantly influenced by the socioeconomic status and social network structure of local communities. This situation necessitates the use of a non-stationary model to effectively characterize the intricate patterns of gun violence in Atlanta (Figure 9.3).

9.2.2 Covariate Data

In addition to the gunshot data, publicly available census data provides crucial insights into the spatial covariate effects on the intensity of local gunshot events. This data is derived from the American Community Survey (ACS) 5-year dataset (U.S. Census Bureau, 2023), which includes over 66,000 variables covering a wide range of topics related to the socioeconomic characteristics of the US population. We select six variables as our spatial covariates, representing demographic, economic, housing, and education figures, as shown in Figure 9.4. Analyzing the relationship between these covariates and gunshot events, we observe that areas with higher socioeconomic status, indicated by variables such as income levels and educational attainment, generally report fewer gunshot incidents. Conversely, regions with higher population density, such as urban centers, experience more frequent gunshot events. These observations suggest a potential relationship between the distribution of gunshot events and the socioeconomic characteristics of a neighborhood, as captured by our spatial covariates, supporting the assumptions of our model.

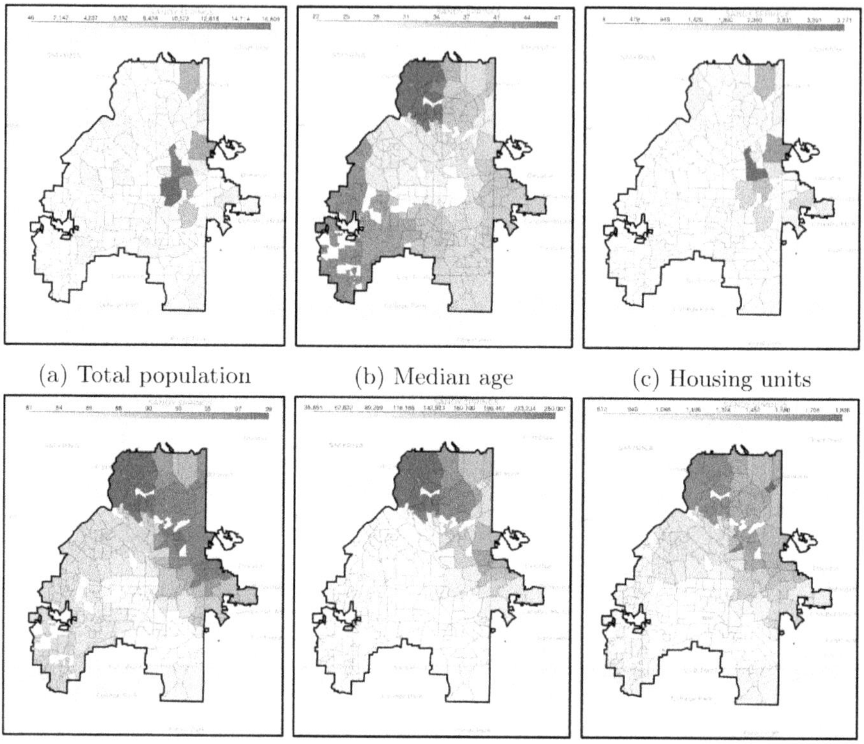

(a) Total population (b) Median age (c) Housing units

(d) High school percentage (e) Median income (f) Median rent

FIGURE 9.4
Six Covariates Were Collected for Each Neighborhood to Study Their Effect on Local Gunshot Event Intensity, Using Values Reported in 2021. (a) Total population. (b) Median age of the population. (c) Occupied housing units. (d) Percentage of the population over 25 years old who have completed high school or higher degrees. (e) Median family income in the past 12 months. (f) Median rent of occupied rental units.

9.3 Model

Motivated by the need to model spatio-temporal gunshot data with spatial covariates and to capture the non-stationary nature of gunshot spread, we adopt a point process model with a non-stationary kernel that incorporates spatial covariates for background intensity. In this section, we first revisit the background of self-exciting spatio-temporal point processes and then introduce our approach to modeling gun-related incidents in Atlanta using a spatial covariate-based point process model.

9.3.1 Background

Consider a sequence of n incidents arranged in time order. Denote the i-th incident as a tuple (t_i, s_i), where $t_i \in [0,\ T]$ represents the time of the occurrence of the incident, and the $s_i \in \mathcal{S}$ represents the location of the incident, measured in longitude and latitude coordinates; $\mathcal{S}$ denotes the observation domain, which, for instance, can be the geographical area covered by the city of Atlanta. The entire event sequence is given by

$$(t_1, s_1), (t_2, s_2), \ldots, (t_n, s_n). \tag{9.1}$$

Spatio-temporal point processes (STPPs) (Reinhart, 2018) have been widely used to model sequences of discrete events that occur in continuous time and space. Let $\mathcal{H}_T = \{(t_i, s_i)\}_{i=1}^{n}$ denote the entire event sequence in $[0, T]$ and $\mathcal{H}_t = \{(t_i, s_i) \in \mathcal{H} \mid t_i < t\}$ denote the observed history before time t. An STPP is fully characterized by the conditional intensity function

$$\lambda(t, s \mid \mathcal{H}_t) = \lim_{\Delta t \downarrow 0, \Delta s \downarrow 0} \frac{\mathbb{E}\left[\mathbb{N}\left([t, t+\Delta t] \times B(s, \Delta s)\right) \mid \mathcal{H}_t\right]}{|B(s, \Delta s)| \Delta t},$$

where $B(s, \Delta s)$ is a ball centered at $s \in \mathbb{R}^2$ with radius Δs, and the counting measure $\mathbb{N}$ is defined as the number of events occurring in $[t,\ t+\Delta t] \times B(s, \Delta s) \subset \mathbb{R}^3$. Therefore, the conditional intensity function can be interpreted as the rate of the occurrence of future events at time t and location s. Naturally $\lambda(t, s \mid \mathcal{H}_t) \geq 0$ for any arbitrary t and s. In the following, we omit the dependency on history $\mathcal{H}_t$ and use a common shorthand $\lambda(t, s)$.

Hawkes process (Hawkes, 1971) is a type of self-exciting point process that captures the triggering effects between events, assuming that influences from past events are linearly additive to the current event. The conditional intensity for a Hawkes point process takes the form of

$$\lambda(t, s) = \mu(s) + \sum_{(t', s') \in \mathcal{H}_t} k(t, t', s, s'), \tag{9.2}$$

where $\mu(s) > 0$ is a location-dependent background rate of events at s, and $k(t, t', s, s')$ is the so-called influence kernel that captures the influence of a past event happened at (t', s') on the likelihood of event occurrence at a future time t and location s. Typically, one parameterizes the kernel $k(t, t', s, s')$ using a certain model with parameter θ; a further simplification is to decompose the kernel as the product of spatial kernel and temporal kernel, and we will adopt such an approach as well.

Given the conditional intensity function (9.2), the log-likelihood of the observed events $\mathcal{H}_T$ can be written as

$$\ell(\theta \mid \mathcal{H}_T) = \sum_{i=1}^{n} \log \lambda(t_i, s_i) - \int_0^T \int_S \lambda(t,s)\, ds dt. \tag{9.3}$$

This enables the estimation of Hawkes process models by determining the kernel parameters θ using maximum likelihood estimation (MLE), which solves $\max_\theta \ell(\theta \mid \mathcal{H}_T)$. To numerically solve the MLE problem, two common approaches are the EM algorithm and gradient descent (Reinhart, 2018). We adopt the latter as it is more memory-efficient (the EM algorithm requires introducing a number of auxiliary parameters that grow quadratically with the number of events) and discuss it further in Section 9.4.

9.3.2 Non-Stationary Kernel with Deep Neural Networks

For model simplicity and computational efficiency, we adopt the commonly used assumption that the triggering effect of a past event is separable in space and time:

$$k(t, t', s, s') = \nu(t, t') \cdot \upsilon(s, s'), \tag{9.4}$$

where $\nu(t, t')$ and $\upsilon(s, s')$ are kernels that capture the temporal and spatial dependence between two events, respectively. We use a non-stationary kernel $\upsilon(s, s')$ to model the contagion of gunshot events, which is further represented by neural networks. Our kernel allows for the modeling of a heterogeneous triggering effect of gunshot events in different regions.

9.3.2.1 Stationary Temporal Kernel

As the goal focuses on capturing the non-homogenous spatial effect, we will assume a relatively simple temporal shift-invariant kernel with exponential decay in (9.4) which takes the following form

$$\nu(t, t') = C e^{-\frac{1}{2\sigma_0^2}(t-t')^2}, t > t'.$$

Here $C > 0$ is a parameter that controls the magnitude of the kernel, and $\sigma_0 > 0$ is a parameter that controls the decaying rate of the event's temporal influences; the causality requirement is captured through $t > t'$ so that a historical event at time t' impacts the current time t but not vice versa.

9.3.2.2 Non-Stationary Spatial Kernel

We will now explain how to construct spatially non-stationary kernel induced by a mixture of Gaussian kernels. Given two arbitrary locations

$s, s' \in \mathcal{S}$, we define the spatial kernel $v(s,s')$ in (9.4) as an inner product between two feature mappings ϕ_s and $\phi_{s'}$, i.e.,

$$v(s,s') = <\phi_s, \phi_{s'}>, \quad s,s' \in S, \tag{9.5}$$

where the inner product for functions $<f,g>:=\int_{\mathbb{R}^2} f(u)g(u)du$. We represent the feature mapping ϕ_s as a weighted sum of a set of R linearly independent feature functions $\left\{\kappa_s^{(r)} := \kappa^{(r)}(s,\cdot)\right\}_{r=1,\ldots,R}$:

$$\phi_s = \sum_{r=1}^{R} w_s^{(r)} \kappa_s^{(r)}, \tag{9.6}$$

where $\kappa_s^{(r)} : \mathcal{S} \times \mathcal{S} \to \mathbb{R}_+$ is a general kernel function and $w_s^{(r)}$ is the corresponding weight of that feature function at location s. The location-dependent weight satisfies $\sum_{r=1}^{R} w_s^{(r)} = 1$ at any arbitrary location s. Hence the spatial kernel can be re-written as

$$v(s,s') = \sum_{1 \le r_1, r_2 \le R} w_s^{(r_1)} w_{s'}^{(r_2)} <\kappa_s^{(r_1)}, \kappa_{s'}^{(r_2)}>.$$

The rationale of this design is twofold: (a) We use a linear combination of the product of feature functions to enhance the representative power of the spatial kernel. (b) The spatial kernel can also be highly interpretable if $\kappa_s^{(r)}$ takes a specific parametric form.

We will follow an approach originally developed for non-stationary Gaussian kernels in Higdon et al. (1999) and later extended to non-stationary triggering kernels in spatio-temporal Hawkes processes in Zhu et al. (2022a). Choose $\kappa_s^{(r)}$ as a Gaussian kernel function centered at s with covariance matrix Σ_s. This results in a simple kernel form, as it can be shown that, using the parametric form of Gaussian kernel functions as basis functions for the feature, the spatial kernel is given by the following as a mixture of Gaussian components

$$v(s,s') = \sum_{1 \le r_1, r_2 \le R} \frac{w_s^{(r_1)} w_{s'}^{(r_2)}}{2\pi \left|\Sigma_s^{(r_1)} + \Sigma_{s'}^{(r_2)}\right|^{\frac{1}{2}}} \exp\left\{-\frac{1}{2}(s-s')\left(\Sigma_s^{(r_1)} + \Sigma_{s'}^{(r_2)}\right)^{-1}(s-s')\right\} \tag{9.7}$$

and thus such construction can still lead to interpretable spatial kernels.

Now we further specify the kernel-induced feature function $\kappa_s^{(r)}$. We will use a parametric form for the Gaussian kernel, inspired by the non-stationary Gaussian process kernel design in Higdon et al. (1999). Since there exists a one-to-one mapping between a bivariate normal distribution specified by Σ_s and its one standard deviation ellipse. Note that $\kappa_s^{(r)}$ is centered at s, so the ellipse's center is fixed at s. Thus we can specify the ellipse by a pair of foci and its area A. The foci are denoted by

$$\psi_s = (\psi_x(s), \psi_y(s)), -\psi_s = (-\psi_x(s), -\psi_y(s)),$$

where $\psi_s \in \mathbb{R}^2, \psi_x, \psi_y : \mathbb{R}^2 \to \mathbb{R}$. Hence, given ψ_s and A, the corresponding Σ_s can be written as

$$\Sigma_s = \tau^2 \begin{pmatrix} Q + \frac{||\psi_s||^2}{2}\cos 2\alpha & \frac{||\psi_s||^2}{2}\sin 2\alpha \\ \frac{||\psi_s||^2}{2}\sin 2\alpha & Q - \frac{||\psi_s||^2}{2}\cos 2\alpha \end{pmatrix}$$

where

$$Q = \sqrt{4A^2 + ||\psi_s||^4 \pi^2} / 2\pi, \alpha = \tan^{-1}(\psi_y(s)/\psi_x(s)),$$

$\tau > 0$ is a scaling parameter that controls the overall level of the covariance.

9.3.2.3 Neural Network-Based Kernel Representation

We develop a neural network-based representation for the kernel-induced feature function; similar idea has been used in Zhu et al. (2022a, c). Put everything above together: to evaluate $v(s, s')$ at any pair of locations s and s', through (9.5), we simply need the location-dependent weights $w_s^{(r)}$ and kernel function $\kappa_s^{(r)}$. The kernel function can be specified using location-dependent focal points ψ_s and location-independent area A. To this end, we parameterize the spatial kernel using neural networks that map the location to the focus points (psi_s) and the weights in (0,1), such that the input of the neural network is the two-dimensional location vector s, and the output is the concatenation of the corresponding focus point ψ_s and its weight w_s. We do this using a fully connected multi-layer neural network. The neural network-based representation will induce smoothly changing spatial kernel with location, as illustrated in Figure 9.5.

In implementation, we adopt the same network architecture for all R kernel-induced feature functions, each modeled as a fully connected network with two hidden layers, where each hidden layer uses the Softplus activation

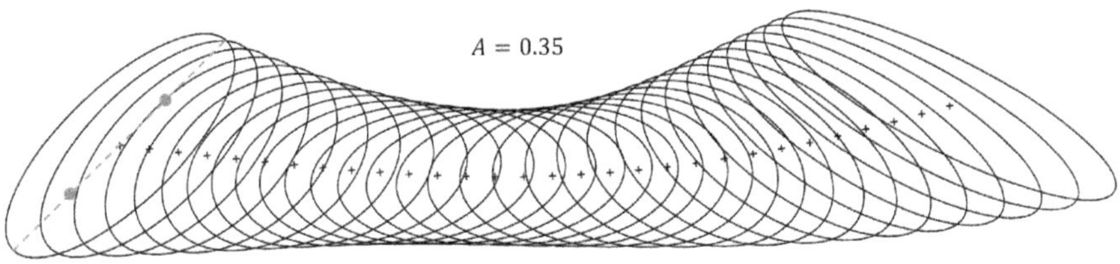

FIGURE 9.5
Illustration of Ellipses Representing the Covariance Matrix of the Spatially Non-stationary Kernel $\kappa_s^{(r)}$ for a Fixed Area $A = 0.35$, with the Two Blue Dots Indicating the Foci.

function $f(x) = \log(1 + e^x)$. For computational simplicity, we assume that the area of the ellipse remains constant at A across all locations and treat A as a hyperparameter. Numerically, we demonstrate that in our setting, neural networks offer a flexible representation of the covariance and the corresponding kernel-induced feature function due to their strong representation capability.

9.3.3 Modeling Spatial Covariates

Spatial covariates, such as demographic variables, also contribute to the spatial heterogeneity of gunshot occurrences. To characterize the exogenous effect of spatial covariates on the gunshot event intensity over space, we first discretize the entire observation domain S into multiple subregions $\{S_1, S_2, \ldots, S_J\}$, and then collect L spatial covariates for each subregion S_j, denoted by $\{x_{j1}, \ldots, x_{jL}\}$, based on which we can establish the connection between the spatial covariates and the gunshot event intensity.

Specifically, we incorporate the covariate effect by introducing a location-dependent background intensity in the conditional intensity function:

$$\mu(s) = \mu_0 + \exp\left\{ \sum_{l=1}^{L} \sum_{j \in N(s)} \gamma_l x_{jl} \cdot \left(\frac{e^{-\alpha d(s, c_j)}}{\sum_{j' \in N(s)} e^{-\alpha d(s, c_{j'})}} \right) \right\}. \tag{9.8}$$

Here, we assume that the background intensity $\mu(s)$ is time-invariant and use a weighted sum of the spatial covariates to characterize their exogenous effect on the gunshot event intensity. The bandwidth parameter $\alpha > 0$ controls the magnitude of the covariate effect. The c_j represents the geographic centroid of the subregion S_j, and $d(s, c_j)$ measures the Euclidean distance between location s and c_j. We only consider the effects of covariates in the subregion that the location s falls in and other subregions that are geographically adjacent to it, by summing over the corresponding index set $N(s)$.

Each covariate may positively or negatively contribute to event intensity. Thus, we introduce a coefficient γ_l for the l-th covariate to indicate such a correlation and its magnitude, which is learned from data. The exponential function outside the weighted sum ensures the non-negativity of the intensity, and $\mu_0 > 0$ captures the endogenous intensity of gunshots. The weighted sum of the spatial covariates allows for smooth modeling of the exogenous covariate effect on the gunshot events.

9.4 Model Fitting

The learning of our model requires the estimation of parameters in deep neural networks. The fitting of the Hawkes process is discussed in another chapter in this book, such as Holbrook et al. We solve the optimization problem of maximum likelihood estimation (MLE) using gradient descent (GD), which has proven to be effective for model estimation when a large number of parameters are involved (Reinhart, 2018). The learning rate is set to be 0.1, and the learning process continues until convergence.

Two major challenges arise during the evaluation of the log-likelihood function (9.3). These challenges are typical in fitting spatio-temporal point process models using maximum likelihood, particularly in dealing with the double integral of the conditional intensity function in (9.3). Here, the double integral can be re-written as

$$\int_0^T \int_{\mathcal{S}} \lambda(t,s)\, ds dt = \mu_0 |\mathcal{S}| T + \int_0^T \int_{\mathcal{S}} \sum_{(t',s') \in \mathcal{H}_t} k(t,t',s,s')\, ds dt$$

$$+ \int_0^T \int_{\mathcal{S}} \exp\left\{ \sum_{l=1}^{L} \sum_{j \in N(s)} \gamma_l x_{jl} \cdot \left(\frac{e^{-\alpha d(s,c_j)}}{\sum_{j' \in N(s)} e^{-\alpha d(s,c_{j'})}} \right) \right\} ds dt. \tag{9.9}$$

The first challenge lies in the integral of the influence kernel (the second term), which involves the spatio-temporal integral of the summation of the effects from historical events, leading to a computational complexity of $\mathcal{O}(n^3)$ with a total of n events in the event series. Reducing this computational cost is crucial, as Hawkes process models – especially those used for public health crises such as gun violence-often require scalable methods for likelihood evaluation and parameter inference. One example (Holbrook et al., 2022) includes the leverage large-scale parallel implementations to efficiently compute the log-likelihood gradient. We adopt an approximation strategy similar to that used in Dong et al. (2023c) – computing the integral of influence kernel with a complexity of only $\mathcal{O}(n)$. Specifically, we assume that the distance between two focus points of an ellipse (ψ_s and $-\psi_s$) at an arbitrary location s is bounded by a threshold $2c$, for a large distance between focus points can lead to an overstretched ellipse, which is unrealistic in practice. Then, by approximating the kernel-induced feature function κ_s using a Gaussian function with an isotropic one-standard-deviation ellipse centered at s of area A, the integral of the influence kernel can be approximated by

$$\int_0^T \int_S \sum_{(t',s')\in\mathcal{H}_t} k(t,t',s,s')dtds = (1+\in)\left\{\sqrt{2\pi}C\sigma_0 \sum_{i=1}^{n}\left[h\left(\frac{T-t_i}{\sigma_0}\right)-\frac{1}{2}\right]\right\},$$

where $\in$ represents the relative error of the integral approximation, and h is the density function of standard Gaussian distribution. The C and σ_0 are the parameters in the temporal kernel. We note that the relative error is bounded by $|\in| < \max\{U-1,1-1/U\}$, where $U = \left(\sqrt{4A^2+c^4\pi^2}+c^2\pi\right)/2A$. In our experiments, we choose $A = 0.35, c = 0.1$ to control the relative error under 0.05.

Another challenge arises from the integral of the covariate effect. We achieve efficient computation of this integral through a similarly tailored numerical integration strategy. Denoting the entire exponential function as $g(s)$, we overlay a sufficiently dense grid $\mathcal{U} \in \mathcal{S}$ over the space. The function $g(s)$ is then evaluated at each grid point in $\mathcal{U}$, allowing the integral to be approximated by

$$\int_0^T \int_S \exp\left\{\sum_{l=1}^{L}\sum_{j\in N(s)} \gamma_l x_{jl}\cdot\left(\frac{e^{-\alpha d(s,c_j)}}{\sum_{j'\in N(s)} e^{-\alpha d(s,c_{j'})}}\right)\right\}dsdt \triangleq \int_0^T \int_S g(s)dsdt \approx T\frac{|\mathcal{S}|}{|\mathcal{U}|}\sum_{u\in\mathcal{U}} g(u).$$

Note that the $\left\{N(u),\{c_j,d(u,c_j)\}_{j\in N(u)}\right\}_{u\in\mathcal{U}}$ can be pre-computed once the grid $\mathcal{U}$ is given. Therefore, we only need to plug in the values of $\{\gamma_l\}_{l\in L}$ every time the parameters are updated during the GD process.

When fitting the model with spatial covariates x_{jl}, we standardize the data by subtracting the mean from each covariate and dividing by its respective standard deviation. In our dataset, all spatial covariates are numerical. This standardization ensures that each covariate is "comparable" in magnitude.

9.5 Results

Below, we present the numerical results of fitting our model to the Atlanta gunshot and census data. Our analysis covers several aspects: in-sample estimation to validate the model's fit to the data, out-of-sample predictions to assess its forecasting accuracy, and visualization of the learned kernels to highlight the model's interpretability.

9.5.1 In-Sample Estimation

We fit the model to two-year gunshot data and analyze its performance. To demonstrate the effectiveness of our model, we perform in-sample predictions for the number of gunshot events in different regions. The in-sample prediction is conducted as follows: given a time interval denoted as $[t_1, t_2]$, we input the training data into the fitted model and calculate the integral of the conditional intensity function over $[t_1, t_2]$ to estimate the number of gunshot events within that time interval. In our evaluation, we estimate the monthly event numbers using our model, including both the total number of gunshot events within the city of Atlanta and the number of events in each neighborhood.

Figure 9.6 visualizes the in-sample estimations produced by our model and compares them with the actual number of reported gunshot events. Our model excels in accurately estimating the number of gunshot events at both the city and neighborhood levels, indicating a good fit to the data. Despite variations in event numbers across different times and neighborhoods, our model effectively captures the spatio-temporal dynamics. We also report performance metrics, including the mean absolute error (MAE) and mean relative error (MRE), for the in-sample estimation of the total event numbers

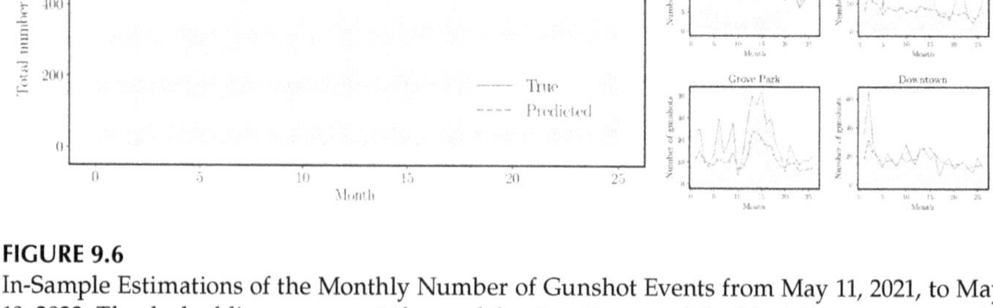

FIGURE 9.6
In-Sample Estimations of the Monthly Number of Gunshot Events from May 11, 2021, to May 10, 2023. The dashed lines represent the model estimations, and the black lines represent the true observations. Left panel: monthly estimation of the total number of gunshots in the city of Atlanta. Right six panels: monthly estimations in the top six neighborhoods with the highest number of gunshots. The mean absolute error (mean relative error) of the estimations for total event numbers is 36.919 (5.64%).

in Atlanta, which indicate reasonably good performance, with a relative error of 5.64%.

9.5.2 Interpretable Spatial Kernel

Our model also demonstrates strong explanatory power through the learned non-stationary spatial kernel. Recall that each $\kappa_s^{(r)}$ is a Gaussian function centered at s with a spatially varying covariance matrix. By interpreting the covariance matrix, we can discover to some extent the underlying spatio-temporal dynamics of the contagious effect of gunshot events. In Figure 9.7, we present three learned feature functions that compose our spatial kernel by visualizing the covariance matrix at different locations. Specifically, we sample 200 locations across the space and draw a line segment at each location, connecting the two focal points of the ellipse corresponding to the covariance matrix. The angle and length of each red line can be interpreted as the direction and strength of the gunshot contagious effect at that specific location.

We observe three distinct modes of the gunshot contagious effect, as indicated by each feature function. The spatial heterogeneity of this effect is clearly evident, underscoring the necessity of our non-stationary kernel. Moreover, detailed characteristics of the gunshot contagious effect in Atlanta can be discerned. For instance, gunshot events in the southwestern area tend to have a broader impact on nearby neighborhoods compared to those in the

(a) $\kappa_s^{(1)}$ (b) $\kappa_s^{(2)}$ (c) $\kappa_s^{(3)}$

FIGURE 9.7
Three Distinctive Modes of the Spatial Kernel for Gunshot Events Are Learned from the Data, as Indicated by the Three Learned Feature Functions $\kappa_s^{(1)}$, $\kappa_s^{(2)}$, and $\kappa_s^{(3)}$. In this visualization, each dashed red line segment connects two focus points of the ellipse corresponding to the Gaussian function $\kappa_s^{(r)}$ at location s (200 line segments are sampled over space in each panel). The shaded area shows the magnitude of the weight $w_s^{(r)}$ over space, with darker colors indicating larger weights.

TABLE 9.1

Estimated $\{\gamma_l\}_{l=1}^{L}$ of Spatial Covariates

Covariate l	γ_l	**Covariate** l	γ_l
Total population	2.2851	High school percentage	–1.8666
Median age	–1.9211	Median income	–3.8588
Housing units	–0.5737	Median rent	–0.3127

northeastern part of Atlanta, providing valuable insights for local practitioners to focus more attention on gunshot risk in those areas.

9.5.3 Estimated Covariate Effect

Another feature of our model is its capability to estimate the covariate effects, which can be analyzed through the fitted coefficients $\{\gamma_l\}_{l=1}^{L}$. Table 9.1 presents the fitted $\{\gamma_l\}_{l=1}^{L}$ alongside the corresponding covariate names. The sign of each coefficient provides crucial insights into the interpretation of the covariate effect. For example, while a larger population size in a neighborhood tends to increase the likelihood of observing gunshot events, other covariates, such as the high school graduation rate and median income, exhibit a negative (inhibitive) effect on gunshot intensity. These inhibitive covariate effects align with the understanding that gun violence is less prevalent among wealthier and more educated populations. The magnitude of the coefficients also highlights the importance of considering these covariates when evaluating gun violence risk. The results indicate that the covariates of total population and median income have the largest effect magnitudes. We would like to emphasize that we can compare the magnitude of the coefficients, because we standardize the covariates before being input into the model.

9.5.4 Out-of-Sample Prediction

To assess the model's ability to forecast future events, we use the fitted model to perform out-of-sample predictions for the conditional intensity function from May 11, 2023, to June 24, 2023, and compare these predictions with the actual gunshot data collected during this period. In out-of-sample prediction, the predicted conditional intensity function can be used to forecast gunshot hotspots. The conditional intensity function at any given time t^* during the testing period can be evaluated by inputting the data prior to t^* into the learned model. Figure 9.8 displays the predicted conditional intensity over space at three different weeks during the testing period. We observe that the predicted intensity from our model aligns well with the spatial distribution of gunshot occurrences, with higher values predicted in areas that subsequently experienced more gunshot events in the following days.

(a) May 22, 2023 (b) June 09, 2023 (c) June 20, 2023

FIGURE 9.8
Predicted Conditional Intensity Function over Space at Three Testing Weeks. The color depth represents the magnitude of the predicted intensity. The dots are reported gunshot events during that week. Note that these events are testing data that are not included for predicting the conditional intensity, and our predictions align well with these events.

A more quantitative way to assess the model's predictive ability is to compute the predicted number of gunshot events via the conditional intensity function and evaluate the prediction error using actual observations. We perform out-of-sample predictions for the number of gunshot events on a weekly basis. For a given time t^*, the predicted event number in a neighborhood during the following week is obtained by integrating the conditional intensity function over time and the neighborhood, similar to the in-sample estimations. The MAE between the predictions and the true observations is then computed to indicate the model's predictive performance. We also perform out-of-sample predictions using three baseline models over the same period and compare their performance with our model: (a) a standard persistent prediction model (Persistent) that uses the event number from the previous week as the prediction for the next week; (b) an ARIMA time series model (ARIMA); and (c) an Epidemic-type Aftershock Sequence (ETAS) model (Ogata, 1988) with a stationary diffusion kernel in Euclidean space.

Table 9.2 reports three types of MAEs as the final evaluation metrics: MAE (frequent) represents the errors of event number predictions averaged over

TABLE 9.2
Out-of-Sample Prediction Performance by Different Models

Model	MAE (Rare)	MAE (Frequent)	MAE (Total)
Persistent	0.337	2.116	0.679
ARIMA	0.354	1.763	0.625
ETAS	$0.392_{(0.002)}$	$2.827_{(0.024)}$	$0.792_{(0.008)}$
Our model	$\mathbf{0.305}_{(0.003)}$	$\mathbf{1.694}_{(0.018)}$	$0.572_{(0.006)}$

the top 20% of neighborhoods with the highest number of gunshot events, MAE (rare) is the average prediction error over the remaining 80% of neighborhoods, and MAE (total) is the average across all neighborhoods. Our model consistently achieves the lowest MAE across all three metrics, demonstrating superior performance compared to other baselines in forecasting gunshot events across neighborhoods with varying characteristics. While ARIMA exhibits performance metrics closest to our model, it is a time series model designed for predicting event numbers and is not intended for handling discrete spatio-temporal events or providing insights into the underlying event dynamics.

9.6 Discussions

We have presented a study using a novel point process model with a nonstationary spatial kernel to analyze gunshot event data in Atlanta. The nonstationary spatial kernel enables a more accurate representation of the complex contagious effects of gunshot events across space, as traditional models often fail to capture these dynamics with stationary spatial interactions. A heterogeneous background intensity that incorporates the effects of census factors accounts for the inherent variability in gunshot events across different communities. By integrating these elements, our model demonstrates superior performance compared to baselines, such as persistent prediction approaches and ETAS models with stationary kernels, in predicting the number and distribution of future gunshot events. The ability to outperform these baselines confirms the presence of intricate underlying dynamics in gunshot event data that are neither random nor simple repetitions of previous patterns. The performance gain of our model over ETAS (which uses a homogeneous kernel) underscores the necessity of adopting a non-stationary spatial kernel to capture the heterogeneous contagious effects of gunshot events across space.

Looking ahead, several avenues for further research are worth exploring: Developing a statistical framework to quantify the significance of the covariate effects (*e.g.*, p-values) on gunshot events, which has yet to be developed. Such significance levels are crucial for confirming the predictive validity of the covariates and for informing policy formulation. Additionally, incorporating a non-stationary temporal kernel could help capture broader temporal trends and shifts in gunshot event patterns over time, potentially related to holiday effects or changes in law enforcement tactics. Furthermore, simulation tools for point process models could be employed to distinguish the heterogeneity of events arising from covariate effects versus contagious phenomena. Such differentiation could aid in developing more effective policies to prevent gun-related incidents.

Acknowledgment

This work is partially supported by NSF grant DMS-2134037, the Atlanta Police Foundation, and the Coca-Cola Foundation. We would like to thank Giuliano Tissot for the helpful discussions.

Bibliography

Braga, A. A., Papachristos, A. V., and Hureau, D. M. (2010). The concentration and stability of gun violence at micro places in Boston, 1980–2008. *Journal of Quantitative Criminology*, 26:33–53.

Cohen, J. and Tita, G. (1999). Diffusion in homicide: Exploring a general method for detecting spatial diffusion processes. *Journal of Quantitative Criminology*, 15:451–493.

De Tarde, G. (1903). *The Laws of Imitation*. H. Holt.

Dong, Z., Cheng, X., and Xie, Y. (2023a). Spatio-temporal point processes with deep non-stationary kernels. In *The Eleventh International Conference on Learning Representations*.

Dong, Z., Repasky, M., Cheng, X., and Xie, Y. (2023b). Deep graph kernel point processes. In *Temporal Graph Learning Workshop@ NeurIPS 2023*.

Dong, Z., Zhu, S., Xie, Y., Mateu, J., and Rodríguez-Cortés, F. J. (2023c). Non-stationary spatio-temporal point process modeling for high-resolution COVID-19 data. *Journal of the Royal Statistical Society Series C: Applied Statistics*, 72(2):368–386.

Fagan, J., Wilkinson, D. L., and Davies, G. (2007). Social contagion of violence. *In D. J. Flannery, A. T. Vazsonyi, and I. D. Waldman (Eds.), The Cambridge Handbook of Violent Behavior and Aggression*. Cambridge University Press, 688–724.

Green, B., Horel, T., and Papachristos, A. V. (2017). Modeling contagion through social networks to explain and predict gunshot violence in Chicago, 2006 to 2014. *JAMA Internal Medicine*, 177(3):326–333.

Hawkes, A. G. (1971). Spectra of some self-exciting and mutually exciting point processes. *Biometrika*, 58(1):83–90.

Higdon, D., Swall, J., and Kern, J. (1999). Non-stationary spatial modeling. *Bayesian Statistics*, 6:761–768.

Holbrook, A. J., Ji, X., and Suchard, M. A. (2022). Bayesian mitigation of spatial coarsening for a hawkes model applied to gunfire, wildfire and viral contagion. *The Annals of Applied Statistics*, 16(1):573–595.

Johnson, S. D. (2008). Repeat burglary victimisation: A tale of two theories. *Journal of Experimental Criminology*, 4:215–240.

Kravitz-Wirtz, N., Bruns, A., Aubel, A. J., Zhang, X., and Buggs, S. A. (2022). Inequities in community exposure to deadly gun violence by race/ethnicity, poverty, and neighborhood disadvantage among youth in large US cities. *Journal of Urban Health*, 99(4):610–625.

Lewis, E., Mohler, G., Brantingham, P. J., and Bertozzi, A. L. (2012). Self-exciting point process models of civilian deaths in Iraq. *Security Journal*, 25:244–264.

Loeffler, C. and Flaxman, S. (2018). Is gun violence contagious? A spatiotemporal test. *Journal of Quantitative Criminology*, 34:999–1017.
Loftin, C. (1986). Assaultive violence as a contagious social process. *Bulletin of the New York Academy of Medicine*, 62(5):550.
Messner, S. F., Anselin, L., Baller, R. D., Hawkins, D. F., Deane, G., and Tolnay, S. E. (1999). The spatial patterning of county homicide rates: An application of exploratory spatial data analysis. *Journal of Quantitative Criminology*, 15:423–450.
Mohler, G. (2013). Modeling and estimation of multi-source clustering in crime and security data. *The Annals of Applied Statistics*, 7:1525–1539.
Mohler, G. (2014). Marked point process hotspot maps for homicide and gun crime prediction in Chicago. *International Journal of Forecasting*, 30(3):491–497.
Mohler, G. O., Short, M. B., Brantingham, P. J., Schoenberg, F. P., and Tita, G. E. (2011). Self-exciting point process modeling of crime. *Journal of the American Statistical Association*, 106(493):100–108.
Morenoff, J. D., Sampson, R. J., and Raudenbush, S. W. (2001). Neighborhood inequality, collective efficacy, and the spatial dynamics of urban violence. *Criminology*, 39(3):517–558.
Ogata, Y. (1988). Statistical models for earthquake occurrences and residual analysis for point processes. *Journal of the American Statistical Association*, 83(401):9–27.
Papachristos, A. V. (2009). Murder by structure: Dominance relations and the social structure of gang homicide. *American Journal of Sociology*, 115(1):74–128.
Ratcliffe, J. H. and Rengert, G. F. (2008). Near-repeat patterns in Philadelphia shootings. *Security Journal*, 21:58–76.
Reinhart, A. (2018). A review of self-exciting spatio-temporal point processes and their applications. *Statistical Science*, 33(3):299–318.
Reinhart, A. and Greenhouse, J. (2018). Self-exciting point processes with spatial covariates: Modelling the dynamics of crime. *Journal of the Royal Statistical Society Series C: Applied Statistics*, 67(5):1305–1329.
Rosenfeld, R., Bray, T. M., and Egley, A. (1999). Facilitating violence: A comparison of gang-motivated, gang-affiliated, and nongang youth homicides. *Journal of Quantitative Criminology*, 15:495–516.
Roth, J. A. and Reiss Jr, A. J. (1993). *Understanding and Preventing Violence: Volume 1*. National Academies Press.
Short, M. B., Brantingham, P. J., Bertozzi, A. L., and Tita, G. E. (2010). Dissipation and displacement of hotspots in reaction-diffusion models of crime. *Proceedings of the National Academy of Sciences*, 107(9):3961–3965.
Short, M. B., Mohler, G. O., Brantingham, P. J., and Tita, G. E. (2014). Gang rivalry dynamics via coupled point process networks. *Discrete and Continuous Dynamical Systems - Series B*, 19(5):1459–1477.
Skogan, W. G., Hartnett, S. M., Bump, N., and Dubois, J. (2008). *Evaluation of CeaseFire-Chicago*. Chicago: Northwestern University.
Taylor, R. M., Simon, M. A., and Patel, D. M. (2013). *Contagion of Violence: Workshop Summary*. National Academies Press.
U.S. Census Bureau. (2023). American Community Survey 5-Year Data (2009–2022), https://www.census.gov/data/developers/data-sets/acs-5year.html.
Webster, D. W., Whitehill, J. M., Vernick, J. S., and Curriero, F. C. (2013). Effects of Baltimore's safe streets program on gun violence: A replication of Chicago's ceasefire program. *Journal of Urban Health*, 90:27–40.

Xu, J. and Griffiths, E. (2017). Shooting on the street: Measuring the spatial influence of physical features on gun violence in a bounded street network. *Journal of Quantitative Criminology*, 33:237–253.

Zhu, S., Bukharin, A., Xie, L., Yamin, K., Yang, S., Keskinocak, P., and Xie, Y. (2022a). Early detection of COVID-19 hotspots using spatio-temporal data. *IEEE Journal of Selected Topics in Signal Processing*, 16(2):250–260.

Zhu, S., Li, S., Peng, Z., and Xie, Y. (2022b). Imitation learning of neural spatio-temporal point processes. *IEEE Transactions on Knowledge and Data Engineering*, 34(11):5391–5402.

Zhu, S., Wang, H., Dong, Z., Cheng, X., and Xie, Y. (2022c). Neural spectral marked point processes. In *The Tenth International Conference on Learning Representations*.

Zhu, S. and Xie, Y. (2022). Spatiotemporal-textual point processes for crime linkage detection. *The Annals of Applied Statistics*, 16(2):1151–1170.

Zhuang, J. and Mateu, J. (2019). A semiparametric spatiotemporal Hawkes-type point process model with periodic background for crime data. *Journal of the Royal Statistical Society: Series A (Statistics in Society)*, 182(3):919–942.

10

Changes in the Reproduction Number of Mass Shootings in the United States Following the COVID-19 Pandemic

Sarah Oliver, Sarah Penrice, P. Jeffrey Brantingham, Fred Morstatter, George Mohler, Alex Knorre

10.1 Introduction

Since the start of the COVID-19 pandemic in the United States in March 2020, rates of violence have been impacted by stay-at-home orders, changes in mobility, and social unrest.[1] Domestic violence increased after the start of the pandemic, a result observed across many cities, states, and countries (Piquero et al. 2021). Gun violence has also increased since 2020 (Ssentongo et al. 2021; Kim 2022), and much of the increase has been concentrated in small "hotspot" areas of US cities (MacDonald, Mohler, and Brantingham 2022). While some of the increase in gun violence is attributable to overall increases in random, endemic violence, a portion of the increase can be explained by an increase in the contagiousness of shootings in general (Brantingham et al. 2021).

Mass shootings have also increased since the start of the pandemic (Schildkraut and Turanovic 2022; Peña and Jena 2021). In Figure 10.1, we display monthly mass shooting counts based on data from the Gun Violence Archive (GVA). There were on average 34 mass shootings per month prior to March 2020, in comparison to over 54 shootings per month from March 2020 to February 2023.

One potential explanation for the increase in mass shootings is that the reproduction number, the expected number of secondary mass shootings triggered by any one prior mass shooting, has increased. Several recent studies have modeled the contagiousness of mass shootings in the United States. In Towers et al. (2015), mass killings involving a firearm were estimated to

 DOI: 10.1201/9781003430773-11

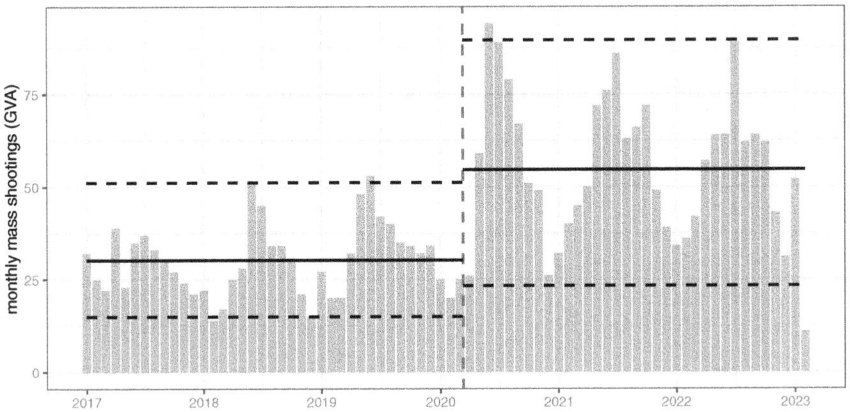

FIGURE 10.1
Monthly Mass Shootings According to the Gun Violence Archive, along with Mean and 95% Percentile of Monthly Shootings Before and After the Start of the Pandemic. Dashed vertical line indicates March 11, 2020.

have a reproduction number of 0.30, and school shootings were estimated to have a reproduction number of 0.22. In Boyd and Molyneux (2021), a Hawkes process was used to estimate the reproduction number of mass shootings. Using GVA data, the authors estimated a reproduction number of 0.66 and an increased risk that lasted for several weeks after each mass shooting event. In Liu, Dong, and Qiu (2022), a Hawkes process was again used to estimate the impact of Twitter posts on subsequent mass shootings.

In the present study, we investigate the extent to which the reproduction number (contagiousness) of mass shootings has increased since the start of the COVID-19 pandemic. For this purpose, we estimate a discontinuity in a Hawkes process model of the reproduction number of mass shootings. We find that the reproduction number increased after March 2020, a result that is consistent across three different mass shooting databases (though the amount of increase in the reproduction number varies across datasets from 0.096 to 0.337).

These findings contribute to the emerging body of knowledge on the contagiousness of mass shootings, particularly during the time of a major social disruption related to the COVID-19 pandemic and social unrest. By comparing the contagiousness of mass shootings before and after the introduction of the national COVID-19-related lockdown, we document the increase in the reproduction number which can be solely attributed to the societal shifts during this time; importantly, this increase is robust to the choice of dataset of mass shootings for modeling.

10.2 Background

Since the start of the COVID-19 pandemic in the United States in March 2020, rates of crime and violence have been impacted by stay-at-home orders, changes in mobility, and social unrest.[1] These changes were puzzling for many scholars of crime. While the majority of crimes went down due to changes in routine activities, the shootings and homicides were a notable exception (Abrams 2021; Lopez and Rosenfeld 2021). Mass shootings have also disproportionately increased in 2020, with an estimated 343 excess mass shootings resulting in 217 fatal and 1498 nonfatal additional victims (Peña and Jena 2021).

Although mass shootings comprise approximately 0.3% of all firearm-related deaths in the US (Schildkraut and Turanovic 2022), they gain disproportionate attention from the media. Instant coverage of these events in the national media environment, primarily TV programs and newspapers, is one major channel allowing "copycat" behavior that enables potential shooters to gain fame and attention. Fox et al. (2021) documented the extent of this disproportionate media attention for particular types of mass shootings. Using the data on 97 mass public shootings from 2000 to 2019 and the matched news coverage from Associated Press, they found that shootings are more likely to be covered in the news if they have more victims, are committed in government and religious facilities, and schools by younger (less than 25 years old) offenders, and are driven by racist and terrorist motives.

Mass shootings have also gradually grown in fatality over the last few decades, even before 2020, despite policy efforts and technological advances that can help prevent this kind of violence. Lankford and Silver (2020) show that in the decade of 2010–2019, the share of high-fatality mass shootings involving eight or more fatal victims was 53% compared to less than in all previous years since 1966. A similar increase occurred with the share of public mass shooters who were influenced by previous incidents (50% in 2010–2019 vs 25% in 1966–2009). Other notable changes include extended planning periods and more frequent use of assault weapons and semiautomatic rifles. In a similar vein, Duwe (2020) also reports the gradual growth in severity (measured by the number of victims shot) and incidence of mass public shootings over 2000–2019. Lankford and Silver (2020) offer a compelling interpretation of this change: in the changing social media and news environment, committing these ferocious acts is a quick way to gain fame and attention, particularly when more victims are killed.

Scholars describe this "copycatting" pattern as a specific contagious type of epidemic violence, as opposed to endemic, or "background" violence (Fox, Sanders, et al. 2021). Given the disproportionate media attention to mass violence, particularly to mass shootings and mass shootings in public places with stranger victims, many studies have emerged that attempt to model the contagiousness of mass shootings. They usually employ self-excitatory point process models, known as Hawkes models which enable scholars to disentangle the self-excitatory and background components of the data-generating process, using spatial and temporal proximity of previous events as predictors of the probability of self-excitation. In other words, these models enable scholars to quantitatively assess to what degree a new event will happen if a previous event happened recently and nearby. This approach has also been used to study the contagiousness of "usual" shootings. Using the incident-level data on shootings in Washington D.C., Loeffler and Flaxman (2018) found that regular gun violence is self-excitatory only on a very limited temporal and spatial scale, suggesting that shootings, at least the majority of them, are not exactly contagious.

The studies of pre-pandemic mass violence suggest that epidemic contagion plays an important role in the spread of these acts of violence. Towers et al. (2015) conducted one of the first studies to assess the contagiousness of mass murders. Drawing from the USA Today and the Brady Campaign national datasets on mass killings and school shootings, the authors use the self-excitatory contagion model to estimate that each incident on average leads to 0.22–0.30 new incidents of mass violence. This contagion effect lasts around 13 days. Notably, the authors do not find evidence of spatial clustering which suggests that similar incidents occurring due to the contagion are unlikely to be spatially correlated. Using four datasets from Brady Campaign, Stanford, Mother Jones, and GVA, Boyd and Molyneux (2021) employ the nonparametric Hawkes process to estimate the reproduction number of each new incident to be 0.45 to 0.89. However, their estimates vary depending on the dataset used and are likely to be driven by various criteria for incident classification and different periods covered.

Kissner (2016) analyzed the temporal contagiousness of active shooting incidents. Using a series hazard model with data on 148 active shootings in New York City in 2000–2012, Kissner shows that the mean inter-incident duration is 31 days; however, the risk of a new active shooting situation significantly increases during 2 weeks from the last similar incident.

Lankford and Tomek (2018) assess the social contagion of mass killings. Using the sample of 232 mass killings from 2006 to 2013 and employing a permutation test for dates of these incidents, they argue that, if there were no contagiousness and the incidents were happening randomly in time, the average time between incidents would be approximately 13 days, ruling out short-term epidemic effects of mass killings.

Taken together, these studies suggest that contagiousness is an important observable component of mass shooting violence. However, the increase in mass shootings in 2020 provided a new puzzle for scholars of mass violence, as its mechanisms are not well-understood. The first and the only analysis of the change in mass shootings before, during, and after the pandemic by Schildkraut and Turanovic (2022) focused specifically on the changes in mass shooting rates after the start of the COVID-19 pandemic. They compare 1,067 public and private mass shootings from the Gun Violence Archive in 2019–2021 and offer a nuanced view of the change in mass shootings in 2020. The rate of both public and private shootings had increased right after the start of the national lockdown in March. However, they also document another increase in public mass shootings following the murder of George Floyd on May 25, 2020. The strain accumulated due to pandemic-related anxieties, a decrease in social interactions, and, finally, anger and sadness after Floyd's death might be a driving force behind this spike, especially coupled with the general shift in routine activities, increased alcohol consumption and gun sales (Schildkraut and Turanovic 2022). Similarly, Brantingham et al. (2021) found that, while the majority of regular shootings that occurred in 2020 after the start of the pandemic were spontaneous, the share of contagious-related shootings has also grown from the start of the pandemic.

However, it is not clear how COVID-19 impacted the contagiousness of mass shooting events in the US. This chapter offers new evidence on the self-excitatory component of mass shootings during the major public health disruption to social life and changes in violent crime.

10.3 Data and Methods

10.3.1 Data

The main issue in studying mass shootings is the lack of a unified definition across different datasets compiled by researchers and nongovernmental organizations (Duwe 2020; Fox, Sanders, et al. 2021). Usually, mass murder is a murder of four or more victims killed during 24 hours. The use of firearms for such murders constitutes a mass shooting. If this happens in a public space absenting any other crime, this becomes a mass public shooting. However, these criteria are not always clear when scholars use secondary datasets compiled by other parties. Thus, mass shootings are situated on the continuum between regular shootings and mass public shootings.

Acknowledging this inherent measurement issue, we focus on mass shootings compiled from several different existing datasets. We draw upon three open-source databases to study the contagiousness of mass shootings since the start of the pandemic: the Gun Violence Archive,[2] a not-for-profit organization with the mission of providing the most comprehensive data that can aid in the conversation surrounding gun violence; Mother Jones,[3] an independent American news organization; and USA Today/Associated Press/Northeastern University, a partnership that aims to "represent the most complete tracking of mass murders."[4]

10.3.1.1 Gun Violence Archive

The Gun Violence Archive (GVA) dataset offers an open-source database collected daily from over 7,500 law enforcement, media, government, and commercial sources. All data is verified by GVA researchers and by an automated system. The GVA defines a mass shooting based on the following criteria:

- At least four or more deaths or injuries (excluding the perpetrator).
- All shooting events are carried out in one location (or within close proximity to the initial event).

The GVA provides data from over 6,000 mass shooting events that have taken place since 2014 in the USA. For this study, we limited the data we analyzed to only include mass shooting events that occurred between January 1, 2017, and February 8, 2023, to have similar observation window lengths before and after the start of the COVID-19 pandemic. This subset of the data includes 3,112 mass shooting events. Each event contains information on the date and location of the event, and how many people were injured and killed as a result of the shooting.[5]

10.3.1.2 The Mother Jones

The Mother Jones database includes 138 events from 1982 to 2023 based on the following criteria:

- at least four deaths (excluding the perpetrator) in a single incident,
- carried out by a lone shooter, and
- in a public place.

The database also includes a select number of "spree killings" in which the killings occurred in multiple locations in a short period, that otherwise fit the above criteria. Mother Jones aims to create criteria that differentiate mass shootings from other types of gun violence such as gang violence or armed robbery.

In our data processing, we filtered Mother Jones to include data from 2010 onward, therefore making use of 87 events from 2010 to 2023 in our model. Each event in the dataset includes a unique label, the date and location of the incident, the number of individuals killed, the number of individuals injured, the type of location, the age of the shooter, whether any mental health issues were noted previous to the incident, and if the weapons were legally obtained or not.

10.3.1.3 USA Today

The database created by the partnership between USA Today, the Associated Press, and Northeastern University contains information about 557 mass killing incidents. A mass killing incident is defined as "the intentional killing of four or more victims – excluding the deaths of unborn children and the offender(s) – by any means within a 24-hour period." These events are gathered using the FBI's Supplementary Homicide Reports and through extensive searches via internet search engines. All events are corroborated by multiple sources such as media accounts, court documents, law enforcement records, etc. If an incident was not able to be successfully verified, it was excluded from the final dataset.

Of these 557 mass killing events, we filtered the data to include only events that are deemed fatal mass shootings defined by the following criteria:

- Four or more victim fatalities.
- Most or all victims are killed by gunfire.
- Within a 24-hour period.

We also filtered the data to include only events after 2010, leaving us with 346 mass shooting incidents to examine. All events contain information surrounding the date, location of the event, number of victims killed and injured, number of perpetrators, how victims were killed, and details about the relationship between the perpetrators and victims.

These three databases define a mass shooting in similar ways, but there are key differences to note. The GVA criteria are based solely on numerical criteria in an attempt to eliminate any source of bias that may be introduced when implementing more stringent criteria. On the other hand, Mother Jones filters out gun violence events that can be attributed to motives such as robbery and gang-related crimes. While this limits the dataset to the specific events of interest, this may introduce a bias in the decision-making used to make these distinctions based on motive. In addition, while the Mother Jones and USA Today datasets require four or more fatalities to be classified as a mass shooting, the GVA dataset accounts for victims who were injured but not killed in the criteria of four or more. Similarly, the USA Today data classifies a mass killing to be within a 24-hour period to eliminate "spree killing" events that occur over several days, while the Mother Jones dataset includes data from these events. Lastly, USA Today allows for more than one perpetrator in their definition of a mass killing, while the Mother Jones data specifically requires one sole perpetrator. However, all datasets appear to be comprehensive within their own criteria and meet the vague but universally applicable definition of a mass shooting. We hope that the combination of these three datasets will provide a comprehensive view of the use of our model.

However, it should be noted that while the rate of mass shootings increased during COVID-19 in the GVA and USA Today dataset, the rate decreased according to data provided by Mother Jones (see Figure 10.2). Thus, we use these three different datasets to assess the robustness of changes in the estimated reproduction number of mass shootings to different definitions of mass shooting.

10.3.2 Model

As is done in Boyd and Molyneux (2021), we consider a Hawkes point process model for estimating the reproduction number of mass shootings. The Hawkes process intensity (rate) of events is given by the following equation:

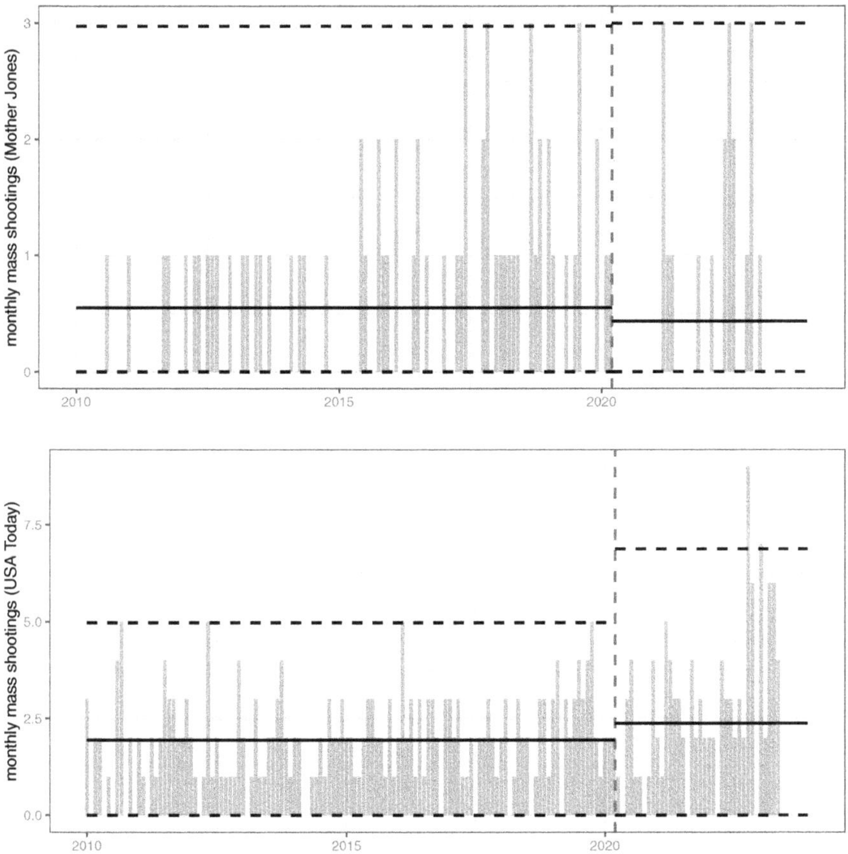

FIGURE 10.2
Monthly Mass Shootings According to Mother Jones (Top) and USA Today (Bottom), along with Mean and 95% Quantile of Monthly Shootings Before and After the Start of the Pandemic. Dashed vertical line indicates March 11, 2020.

$$\lambda(t) = \mu + \sum_{t>t_i} R\gamma e^{-\gamma(t-t_i)}. \tag{10.1}$$

The baseline rate, μ, models "background" or "spontaneous" mass shootings that occur at random according to a Poisson process. In particular, the number of background events, N, is given by a Poisson random variable, $N \sim Pois(\mu t)$. Conditioned on the number of events, these background mass shootings are distributed across a given period of time according to a uniform distribution. The Hawkes process can be viewed as a branching process (Hawkes and Oakes 1974), where the reproduction number R describes the average number of events that are directly caused by an initial event. The parameter γ then determines the timescale between events that are causally linked in the branching process, where γ^{-1} is the average serial interval between a primary and a secondary event. The intensity in Equation (10.1) increases by $R\gamma$ immediately following an event, and thus the probability of another event occurring in the near future is increased.

To analyze the effects of the pandemic on mass shootings, we alter the model to allow for the reproduction number and baseline rate of events to change after the start of the pandemic:

$$\lambda(t) = \mu_t + \sum_{t>t_i} R_t\gamma e^{-\gamma(t-t_i)}, \tag{10.2}$$

where

$$\mu_t = \mu_1 \cdot 1\{t < c\} + \mu_2 \cdot 1\{t \ge c\} \tag{10.3}$$

$$R_t = R_1 \cdot 1\{t < c\} + R_2 \cdot 1\{t \ge c\}. \tag{10.4}$$

Thus we add a discontinuity where the reproduction number and baseline rate of mass shootings are given by R_1 and μ_1 before March 11, 2020 (represented in Equation (10.2) by $t < c$), and by R_2 and μ_2 following the start of the pandemic. We perform Bayesian estimation of the model parameters in STAN (Carpenter et al. 2017) using uninformative priors. We specify that R_1 and R_2 have beta (1, 1) priors, as the reproduction number for the point process in Equation (10.2) needs to be between 0 and 1 for the model to be well-posed. For γ, μ_1, and μ_2 we use exponential (0.1) priors.

10.4 Results

We first estimate the model in Equations (10.2)–(10.4) using GVA data from 2017 to 2023, yielding the posterior parameter estimates presented in Table 10.1. We display the posterior mean and 95% range for each of the model

TABLE 10.1

Estimated Parameters (Posterior Mean and 95% Range) for Equation (10.2) Fit to GVA Data

	Mean	95% Range
R_1	0.225	[0.157, 0.286]
R_2	0.417	[0.345, 0.493]
μ_1	0.772	[0.691, 0.855]
μ_2	1.074	[0.944, 1.209]
γ	1.273	[1.024, 1.558]

parameters, and in Figure 10.3, we plot the posterior mean and 95% range of the fitted intensity.

In Table 10.1, we find that the average estimated reproduction number R_1 before the pandemic was 0.225, whereas R_2 after the epidemic was estimated to be 0.417 (almost double). This increase in the reproduction number is statistically significant. The 95% posterior range of the reproduction numbers does not overlap, and the value of R_2 was higher than R_1 in 100% of the posterior samples. According to this model, mass shootings have been more contagious since the start of the COVID-19 pandemic in the US in March 2020.

Similarly, the baseline rate of events after the pandemic, $\mu_2 = 1.074$, increased over the pre-pandemic value, $\mu_1 = 0.772$. This is consistent with recent research that showed both the reproduction number and base rate of urban gun violence increased after the start of the pandemic (Brantingham et al. 2021).

Next we estimate the model in Equation (10.2) on the smaller Mother Jones dataset from 2010 to 2023, yielding the posterior parameter estimates in Table 10.2. The posterior mean and 95% range of the fitted intensity are shown in Figure 10.4.

The change in the posterior mean reproduction number in Table 10.2 shows that the reproduction number $R_2 = 0.439$ after March 2020 is roughly four times the value of $R_1 = 0.102$ before March 2020. However, due to the small size of the dataset, there is a significant overlap of the 95% range of the posteriors of R_1 and R_2, and in only 91.75% of the posterior samples was the value of R_2 higher than R_1.

Whereas the baseline rate of events increased across the GVA dataset, in the Mother Jones data the estimated baseline rate of mass shootings decreased slightly following the start of the pandemic (μ_2 is 0.005 less than μ_1). However, this decrease is not statistically significant, given the overlap of

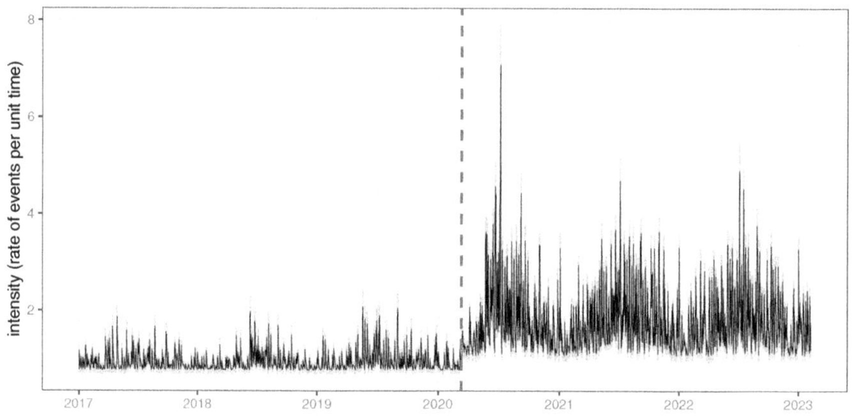

FIGURE 10.3
Posterior 95% Confidence Interval (Gray) and Mean (Black) Intensity of Equation (10.2) Fit to GVA Data. Dashed vertical line indicates March 11, 2020.

TABLE 10.2
Estimated Parameters (Posterior Mean and 95% Range) for Equation (10.2) Fit to Mother Jones Data

	Mean	95% Range
R_1	0.102	[0.005, 0.326]
R_2	0.439	[0.035, 0.892]
μ_1	0.018	[0.012, 0.023]
μ_2	0.0126	[0.004, 0.023]
γ	0.076	[0.011, 0.556]

the two intervals in Table 10.2 and that only 81.5% of the posterior samples of μ_2 were smaller than the corresponding posterior sample for μ_1.

We estimate the model in Equation (10.2) on the USA Today dataset from 2010 to 2023, yielding the posterior parameter estimates in Table 10.3. The posterior mean and 95% range of the fitted intensity are shown in Figure 10.5.

The change in the posterior mean reproduction number in Table 10.3 shows that the reproduction number $R_2 = 0.122$ after March 2020 is over four times the value of $R_1 = 0.027$ before March 2020, similar to the effect observed with the Mother Jones data. However, due to the small size of the dataset, there is a significant overlap of the 95% range of the posteriors of R_1 and R_2. The baseline rate of events is estimated to have increased following the start of

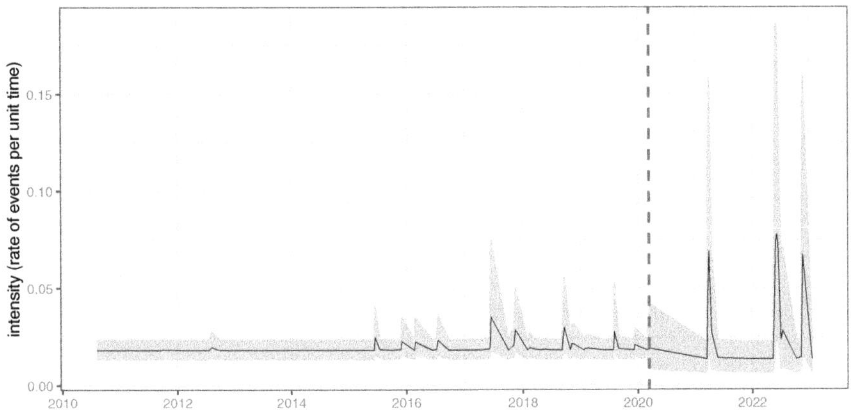

FIGURE 10.4
Posterior 95% Confidence Interval (Gray) and Mean (Black) Intensity of Equation (10.2) Fit to Mother Jones Data. Dashed vertical line indicates March 11, 2020.

TABLE 10.3
Estimated Parameters (Posterior Mean and 95% Range) for Equation (10.2) Fit to USA Today Data

	Mean	95% Range
R_1	0.027	[0.000, 0.147]
R_2	0.122	[0.002, 0.481]
μ_1	0.062	[0.053, 0.072]
μ_2	0.083	[0.0490, 0.108]
γ	1.534	[0.015, 11.953]

the pandemic ($\mu_2 = .083, \mu_1 = .062$). However, this change is not statistically significant.

We also investigate the sensitivity of the change in reproduction number to the cutoff date used in the model. In Tables 10.4, 10.5 and 10.6 we compare the change $R_2 - R_1$ when the COVID-19 cutoff March 2020 is used, to when March 2019 and March 2021 are chosen for the cutoff. In the case of Mother Jones and USA Today, the largest change in $R_2 - R_1$ occurred in March 2020 at the start of the pandemic. However, for the GVA dataset, the change $R_2 - R_1$ is slightly larger in 2019.

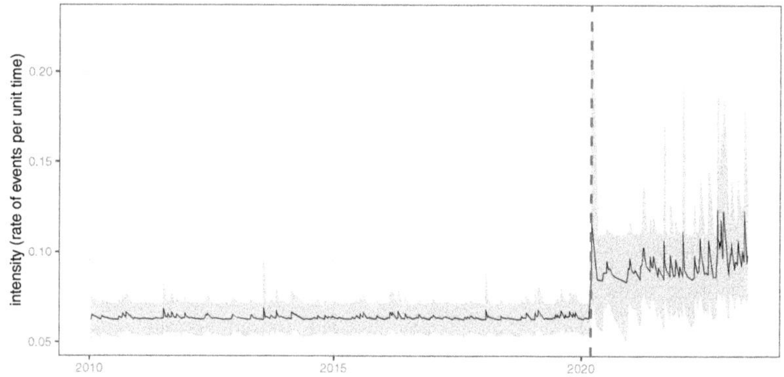

FIGURE 10.5
Posterior 95% Confidence Interval (Gray) and Mean (Black) Intensity of Equation (10.2) Fit to USA Today Data. Dashed vertical line indicates March 11, 2020.

TABLE 10.4

Reproduction Number vs. Cutoff Date for Mother Jones Database

Date Cutoff	R_1	R_2	$R_2 - R_1$
March 2019	0.099	0.330	0.231
March 2020	0.102	0.439	0.337
March 2021	0.135	0.234	0.099

TABLE 10.5

Reproduction Number vs. Cutoff Date for Gun Violence Archive Database

Date Cutoff	R_1	R_2	$R_2 - R_1$
March 2019	0.205	0.433	0.228
March 2020	0.225	0.417	0.192
March 2021	0.398	0.381	−0.017

TABLE 10.6

Reproduction Number vs. Cutoff Date for USA Today Database

Date Cutoff	R_1	R_2	$R_2 - R_1$
March 2019	0.031	0.116	0.085
March 2020	0.027	0.123	0.096
March 2021	0.023	0.090	0.067

10.5 Discussion

We observed an increase in the contagiousness of mass shootings in the United States following the start of the COVID-19 pandemic in March of 2020. This increase was observed in all three data sets: the GVA dataset ($R_1 = 0.225, R_2 = 0.417$), the Mother Jones dataset ($R_1 = 0.102, R_2 = 0.439$), and the USA Today dataset ($R_1 = 0.027, R_2 = 0.123$).

One potential explanation for the contagiousness of mass shootings is that *"stressed individuals may have, consciously or sub-consciously, been inspired to act on previously suppressed urges by exposure to details of similar events"* in the media (Towers et al. 2015). Similar media effects are observed in suicide ideation following portrayals and descriptions of suicide in news and other online media (Gould 2001; Gould, Jamieson, and Romer 2003), which has led the World Health Organization to release media guidelines to prevent suicide (*Preventing suicide: a resource for media professionals, 2023 update 2023*). It is important to emphasize, however, that the contagiousness of mass shootings (and suicide) does not require formal incitement to violence. Rather, the putative mechanism involves individuals fusing ambient hostile rhetoric, amplified by traditional and social media, with their local (private) grievances. This fusion creates the incentive to act independently. Thus, mass shootings appear to be instances of "stochastic terrorism" at the scale of the individual (Amman and Meloy 2022), since there is no formal plan or conspiracy underlying the act. Nevertheless, the role played by exposure to hostile rhetoric is central to the contagion of violence and is visible at the aggregate scale.

However, a general model of "stochastic terrorism" does not explain why there would be a jump in the reproduction number of mass shootings at the onset of the COVID-19 pandemic. We conjecture that something about the pandemic changed the nature of exposure to hostile rhetoric and/or the conditions necessary for individuals to take hostile action in the "real world." Changes in both domains seem plausible. First, the available evidence indicates that time engaged online jumped dramatically with the onset of the pandemic for all age groups (Schmidt et al. 2020; Pandya and Lodha 2021). Increased activity online during the pandemic may have produced both a greater number of initial exposures to hostile rhetoric as well as the opportunity for deeper engagement with such rhetoric for those previously exposed.

Second, the pandemic also impacted offline routine activities by "untethering" people from normative patterns of work, school, and recreation (Mohler et al. 2020). While the range of social distancing measures put into place may have helped "flatten the curve" of COVID-19 transmissions, it may have also created greater freedom or opportunity for individuals to act on their grievances. An individual freed of their obligation to show up (in person) at work or school may have had greater freedom to purchase weapons and put them to use. Of course, the implication is that the mass shooting reproduction number could return to pre-pandemic levels as pandemic conditions continue to recede and behavioral routines return to normal. However, there is no indication as yet that this has occurred. Some changes in online and offline behavior brought on by the pandemic may be permanent.

10.5.1 Limitations

One limitation of the present study is that there is not a comprehensive national database on gun violence, including mass shootings, and that results may depend on the precise definition of mass shooting. We attempted to test the robustness of our results with three datasets; however, two additional datasets used in previous research (the Stanford mass shooting database and a database maintained by the Brady Campaign) have not been up-to-date since the start of the pandemic. Whereas the Mother Jones and USA Today datasets use a more stringent definition of mass shootings, these datasets also have significantly fewer events in comparison to GVA. The reduced dataset size makes it difficult to infer statistically significant trends.

A second limitation of the present study is that we did not include spatial effects in the model. One of the Bradford Hill criteria for causality is consistency (Kucharski 2020), meaning that the results of a study can be reproduced in a different time and place by a different person. Because mass shootings are so much more common in the United States than elsewhere, and state-level models would suffer from small sample sizes, our ability to test consistency is limited. While previous research suggests that the spatial component is not instrumental in the diffusion of mass shootings, further research could investigate spatial clustering of events within regions of the United States and compare mass shooting data across different regions within the United States, particularly since the start of the COVID-19 pandemic.

Dose-response is another factor that suggests causality and could be a line of future research. The current model does not account for how the intensity of mass shootings is related to the number of injuries, fatalities, or the level of media coverage. News coverage could be added into the Hawkes process model in Equation (10.2) as a time-varying covariate (Chiang, Liu, and Mohler 2022). A better understanding of the link between online media and subsequent mass shootings may motivate improved measures taken by news media, such as mandatory disclaimers and disclosure of mental health resources when reporting on suicides in the news.

Notes

1. Authors Sarah Oliver and Sarah Penrice listed alphabetically and contributed equally to the research. This research was supported in part by NSF grants SCC-2125319, ATD-2124313, and AFOSR MURI grant FA9550-22-1-0380.
2. https://www.gunviolencearchive.org

3. https://www.motherjones.com/politics/2012/12/mass-shootings-mother-jones-full-data
4. https://data.world/associatedpress/mass-killings-public
5. https://www.gunviolencearchive.org/about

Bibliography

Abrams, David S. 2021. "COVID and crime: An early empirical look." *Journal of Public Economics* 194 (February 1, 2021): 104344. https://doi.org/10.1016/j.jpubeco.2020.104344.

Amman, Molly, and Reid Meloy. 2022. "Incitement to violence and stochastic terrorism: Legal, academic, and practical parameters for researchers and investigators." *Terrorism and Political Violence*, 1–12. https://doi.org/10.1080/09546553.2022.2143352.

Boyd, Peter, and James Molyneux. 2021. "Assessing the contagiousness of mass shootings with nonparametric Hawkes processes." *PLoS One* 16 (3): e0248437.

Brantingham, P. Jeffrey, Jeremy Carter, John MacDonald, Chris Melde, and George Mohler. 2021. "Is the recent surge in violence in American cities due to contagion?" *Journal of Criminal Justice* 76: 101848.

Carpenter, Bob, Andrew Gelman, Matthew D. Hoffman, Daniel Lee, Ben Goodrich, Michael Betancourt, Marcus Brubaker, Jiqiang Guo, Peter Li, and Allen Riddell. 2017. "Stan: A probabilistic programming language." *Journal of Statistical Software* 76 (1): 1–32.

Chiang, Wen-Hao, Xueying Liu, and George Mohler. 2022. "Hawkes process modeling of COVID-19 with mobility leading indicators and spatial covariates." *International Journal of Forecasting* 38 (2): 505–520.

Duwe, Grant. 2020. "Patterns and prevalence of lethal mass violence." *Criminology & Public Policy* 19 (1): 17–35. https://doi.org/10.1111/1745-9133.12478.

Fox, James Alan, Madison Gerdes, Grant Duwe, and Michael Rocque. 2021. "The news-worthiness of mass public shootings: What factors impact the extent of coverage?" *Homicide Studies* 25, no. 3 (August 1, 2021): 239–255. https://doi.org/10.1177/1088767920974412.

Fox, James Alan, Nathan E. Sanders, FridelEmma E. , Grant Duwe, and Michael Rocque. 2021. "The contagion of mass shootings: The interdependence of large-scale massacres and mass media coverage." *Statistics and Public Policy* 8, no. 1 (January 1, 2021): 53–66. https://doi.org/10.1080/2330443X.2021.1932645.

Gould, Madelyn, Patrick Jamieson, and Daniel Romer. 2003. "Media contagion and suicide among the young." *American Behavioral Scientist* 46 (9): 1269–1284.

Gould, Madelyn S. 2001. "Suicide and the media." *Annals of the New York Academy of Sciences* 932 (1): 200–224.

Hawkes, Alan G., and David Oakes. 1974. "A cluster process representation of a self-exciting process." *Journal of Applied Probability* 11 (3): 493–503.

Kim, Dae-Young. 2022. "The impact of COVID-19 on gun violence across census tracts in NYC." *Homicide Studies* 26 (4): 379–402.

Kissner, Jason. 2016. "Are active shootings temporally contagious? An empirical assessment." *Journal of Police and Criminal Psychology* 31, no. 1 (March 1, 2016): 48–58. https://doi.org/10.1007/s11896-015-9163-8.

Kucharski, Adam. 2020. *The Rules of Contagion: Why Things Spread–And Why They Stop.* Basic Books.

Lankford, Adam, and James Silver. 2020. "Why have public mass shootings become more deadly?" *Criminology & Public Policy* 19 (1): 37–60. https://doi.org/10.1111/1745-9133.12472.

Lankford, Adam, and Sara Tomek. 2018. "Mass killings in the United States from 2006 to 2013: Social contagion or random clusters?" *Suicide and Life-Threatening Behavior* 48 (4): 459–467. https://doi.org/10.1111/sltb.12366.

Liu, Dixizi, Zhijie Sasha Dong, and Guo Qiu. 2022. "Exploring the contagion effect of social media on mass shootings." *Computers & Industrial Engineering* 172: 108565.

Loeffler, Charles, and Seth Flaxman. 2018. "Is gun violence contagious? A spatio-temporal test." *Journal of Quantitative Criminology* 34, no. 4 (December 1, 2018): 999–1017. https://doi.org/10.1007/s10940-017-9363-8.

Lopez, Ernesto, and Richard Rosenfeld. 2021. "Crime, quarantine, and the U.S. coronavirus pandemic." *Criminology & Public Policy* 20 (3): 401–422. https://doi.org/10.1111/1745-9133.12557.

MacDonald, John, George Mohler, and P. Jeffrey Brantingham. 2022. "Association between race, shooting hot spots, and the surge in gun violence during the COVID-19 pandemic in Philadelphia, New York and Los Angeles." *Preventive Medicine* 165: 107241.

Mohler, George, Andrea L. Bertozzi, Jeremy Carter, Martin B. Short, Daniel Sledge, George E. Tita, Craig D. Uchida, and P. Jeffrey Brantingham. 2020. "Impact of social distancing during COVID-19 pandemic on crime in Los Angeles and Indianapolis." *Journal of Criminal Justice* 68: 101692. https://doi.org/10.1016/j.jcrimjus.2020.101692.

Pandya, Apurvakumar, and Pragya Lodha. 2021. "Social connectedness, excessive screen time during COVID-19 and mental health: A review of current evidence." *Frontiers in Human Dynamics* 3. https://doi.org/10.3389/fhumd.2021.684137.

Peña, Pablo A., and Anupam Jena. 2021. "Mass shootings in the US during the COVID-19 pandemic." *JAMA Network Open* 4 (9): e2125388–e2125388.

Piquero, Alex R., Wesley G. Jennings, Erin Jemison, Catherine Kaukinen, and Felicia Marie Knaul. 2021. "Domestic violence during the COVID-19 pandemic-evidence from a systematic review and meta-analysis." *Journal of Criminal Justice* 74: 101806.

Schildkraut, Jaclyn, and Jillian J. Turanovic. 2022. "A new wave of mass shootings? Exploring the potential impact of COVID-19." *Homicide Studies* 26 (4): 362–378.

Schmidt, Steffen CE, Bastian Anedda, Alexander Burchartz, Ana Eichsteller, Simon Kolb, Carina Nigg, Claudia Niessner, Doris Oriwol, Annette Worth, and Alexander Woll. 2020. "Physical activity and screen time of children and adolescents before and during the COVID-19 lockdown in Germany: A natural experiment." *Scientific Reports* 10 (1): 1–12.

Ssentongo, Paddy, Claudio Fronterre, Anna E. Ssentongo, Shailesh Advani, Emily S. Heilbrunn, Joshua P. Hazelton, John S. Oh, Jennifer S. McCall-Hosenfeld, and Vernon M. Chinchilli. 2021. "Gun violence incidence during the COVID-19 pandemic is higher than before the pandemic in the United States." *Scientific Reports* 11 (1): 20654.

Towers, Sherry, Andres Gomez-Lievano, Maryam Khan, Anuj Mubayi, and Carlos Castillo-Chavez. 2015. "Contagion in mass killings and school shootings." *PLoS One* 10 (7): e0117259.

World Health Organization. 2023. *Preventing Suicide: A Resource for Media Professionals, Update.* 2023. https://www.who.int/publications/i/item/9789240076846.

Index

F

G

T

U

V

W

For Product Safety Concerns and Information please contact our EU
representative GPSR@taylorandfrancis.com
Taylor & Francis Verlag GmbH, Kaufingerstraße 24, 80331 München, Germany

www.ingramcontent.com/pod-product-compliance
Ingram Content Group UK Ltd.
Pitfield, Milton Keynes, MK11 3LW, UK
UKHW020158200926
475702UK00004B/511
* 9 7 8 1 0 3 2 5 5 4 4 9 5 *